Combat Modeling

INT. SERIES IN OPERATIONS RESEARCH & MANAGEMENT SCIENCE

Series Editor: Frederick S. Hillier, Stanford University
Special Editorial Consultant: Camille C. Price, Stephen F. Austin State University
Titles with an asterisk (*) were recommended by Dr. Price

Washburn & Kress/ *COMBAT MODELING*
Netessine & Tang/ *CONSUMER-DRIVEN DEMAND AND OPERATIONS MANAGEMENT MODELS: A Systematic Study of Information-Technology-Enabled Sales Mechanisms*
Saaty & Vargas/ *DECISION MAKING WITH THE ANALYTIC NETWORK PROCESS: Economic, Political, Social & Technological Applications w. Benefits, Opportunities, Costs & Risks*
Yu/ *TECHNOLOGY PORTFOLIO PLANNING AND MANAGEMENT: Practical Concepts and Tools*
Kandiller/ *PRINCIPLES OF MATHEMATICS IN OPERATIONS RESEARCH*
Lee & Lee/ *BUILDING SUPPLY CHAIN EXCELLENCE IN EMERGING ECONOMIES*
Weintraub/ *MANAGEMENT OF NATURAL RESOURCES: A Handbook of Operations Research Models, Algorithms, and Implementations*
Hooker/ *INTEGRATED METHODS FOR OPTIMIZATION*
Dawande et al/ *THROUGHPUT OPTIMIZATION IN ROBOTIC CELLS*
Friesz/ *NETWORK SCIENCE, NONLINEAR SCIENCE and INFRASTRUCTURE SYSTEMS*
Cai, Sha & Wong/ *TIME-VARYING NETWORK OPTIMIZATION*
Mamon & Elliott/ *HIDDEN MARKOV MODELS IN FINANCE*
del Castillo/ *PROCESS OPTIMIZATION: A Statistical Approach*
Józefowska/*JUST-IN-TIME SCHEDULING: Models & Algorithms for Computer & Manufacturing Systems*
Yu, Wang & Lai/ *FOREIGN-EXCHANGE-RATE FORECASTING WITH ARTIFICIAL NEURAL NETWORKS*
Beyer et al/ *MARKOVIAN DEMAND INVENTORY MODELS*
Shi & Olafsson/ *NESTED PARTITIONS OPTIMIZATION: Methodology and Applications*
Samaniego/ *SYSTEM SIGNATURES AND THEIR APPLICATIONS IN ENGINEERING RELIABILITY*
Kleijnen / *DESIGN AND ANALYSIS OF SIMULATION EXPERIMENTS*
Førsund/ *HYDROPOWER ECONOMICS*
Kogan & Tapiero/ *SUPPLY CHAIN GAMES: Operations Management and Risk Valuation*
Vanderbei/ *LINEAR PROGRAMMING: Foundations & Extensions, 3rd Edition*
Chhajed & Lowe/ *BUILDING INTUITION: Insights from Basic Operations Mgmt. Models and Principles*
Luenberger & Ye/ *LINEAR AND NONLINEAR PROGRAMMING, 3rd Edition*
Drew et al/ *COMPUTATIONAL PROBABILITY: Algorithms and Applications in the Mathematical Sciences**
Chinneck/ *FEASIBILITY AND INFEASIBILITY IN OPTIMIZATION: Algorithms and Computation Methods*
Tang, Teo & Wei/ *SUPPLY CHAIN ANALYSIS: A Handbook on the Interaction of Information, System and Optimization*
Ozcan/ *HEALTH CARE BENCHMARKING AND PERFORMANCE EVALUATION: An Assessment using Data Envelopment Analysis (DEA)*
Wierenga/ *HANDBOOK OF MARKETING DECISION MODELS*
Agrawal & Smith/*RETAIL SUPPLY CHAIN MANAGEMENT: Quantitative Models and Empirical Studies*
Brill/ *LEVEL CROSSING METHODS IN STOCHASTIC MODELS*
Zsidisin & Ritchie/ *SUPPLY CHAIN RISK: A Handbook of Assessment, Management & Performance*
Matsui/ *MANUFACTURING AND SERVICE ENTERPRISE WITH RISKS: A Stochastic Management Approach*
Zhu/*QUANTITATIVE MODELS FOR PERFORMANCE EVALUATION AND BENCHMARKING: Data Envelopment Analysis with Spreadsheets*
Kubiak/ *PROPORTIONAL OPTIMIZATION AND FAIRNESS**
Bier & Azaiez/ *GAME THEORETIC RISK ANALYSIS OF SECURITY THREATS**

~A list of the early publications in the series is found at the end of the book~

Alan Washburn · Moshe Kress

Combat Modeling

Springer

Alan Washburn
Department of Operations Research
Naval Postgraduate School
Monterey, CA 93943
USA
awashburn@nps.edu

Moshe Kress
Department of Operations Research
Naval Postgraduate School
Monterey, CA 93943
USA
mkress@nps.edu

ISSN 0884-8289
ISBN 978-1-4614-2932-6 e-ISBN 978-1-4419-0790-5
DOI 10.1007/978-1-4419-0790-5
Springer Dordrecht Heidelberg London New York

Printed on acid-free paper

Springer is part of Springer Science+Business Media (www.springer.com)

Preface

This book deals with the processes, methods, and concepts that lie behind modern models of combat. It is intended for readers with at least a scientific bachelor's degree and some background in basic probability concepts. It includes three appendices that address general quantitative methods that are applied throughout the book. Appendix A is a review of probability concepts, which are involved in almost every chapter. A reader who has never studied probability before should consider reading this appendix, at least, before continuing. Appendices B and C are minimal introductions to the topics of optimization and Monte Carlo simulation.

Use will be made throughout this book of Microsoft Excel™ workbooks developed by the authors. Some of these workbooks include VBA (Visual Basic for Applications) code for certain functions or commands. The reader should obtain the zipped folder *CombatModeling1.zip* from the "downloads" link at the url http://faculty.nps.edu/awashburn/ . After obtaining the folder, unzip it and read *Readme.txt* and *Errata1.txt*. As long as proper credit for authorship is given, the workbooks may be freely distributed. An email link for author Washburn will also be found at that url or substitute "mkress" for "awashburn" to reach author Kress.

Feedback will be appreciated. If you find an error, please point it out so we can incorporate it in the errata. If there are subsequent editions of this book, we will take advantage of your comments about organization and utility.

We hope you find our book useful.

Alan Washburn
Operations Research Department
Naval Postgraduate School

Moshe Kress
Operations Research Department
Naval Postgraduate School

Contents

Chapter 1
Generalities and Terminology

I'm no model lady. A model's just an imitation of the real thing.

Mae West

1.1 Introduction

A model is an abstraction of reality. An abstraction can take many forms: an architect might construct a physical miniature model of the building he plans; a CEO of a corporation might use a diagram to present a new business idea, and a physicist might use a set of differential equations to represent some physical phenomenon. The need for models stems from the fact that the real world is too complicated for us to reason about and contains many details that are not necessarily relevant. Our limited intellects permit us to deal only with abstractions that retain the essence of the matter without the distracting details. The miniature model of the building, the diagram, and the set of differential equations are manifestations of these abstractions and henceforth called simply *models*. Thus, models are entities of various types – physical, notional, or mathematical – that share the fact that they represent only an abstraction of a real object or situation. Models are used for reasoning, insight, planning, and prediction. They need to capture the key factors of the object or situation and faithfully represent them so that the models can be utilized effectively.

This book is about combat models, models that describe or represent weapon systems and combat situations. As mentioned above, there are several types. In order to organize our thinking on the matter, it is important to introduce terms by which one type of model can be distinguished from another, as well as terms about how such models are developed, used, and tested. The development of such a lexicon is one object of this chapter (Section 1.2). Another object (Section 1.3) is to discuss certain modeling assumptions that are employed frequently enough to deserve names. It is the authors' contention that these "shortcut" assumptions are dangerous as well as useful and powerful and should therefore be discussed immediately in a book such as this. Section 1.4 summarizes the notational conventions that will be used throughout the book, and Section 1.5 is an overview of subsequent chapters.

A. Washburn, M. Kress, *Combat Modeling*, International Series in Operations Research & Management Science 134, DOI 10.1007/978-1-4419-0790-5_1,

1.2 Classification of Combat Models

Unfortunately, the business of combat modeling has long suffered from a lexicon that often obscures fundamental relationships and parallels. There have been several attempts to remedy this by authorities who announce glossaries of what terms should mean, but none of these have been completely successful. It is more or less like trying to impose standards on (say) French – the people will speak as they wish, despite the best efforts of the authorities. Terms such as "identity simulation" and "agent-based simulation" are regularly used in spite of being missing from such glossaries and seem to ease communication between those in the know. Still, one ought to make an effort to use terms that are in some sense standard. Our approach in this book will roughly follow the glossary defined by the United States Department of Defense (DoD, 1998). Our first, definitive use of terms will be italicized in the following.

The real combat situation will inevitably differ from its model, particularly because of the complexity and uncertainty associated with combat. Judging the extent to which the model agrees with the real world is called *validation,* to be distinguished from judging whether a particular implementation of a model (e.g., a computer program) is correct in the sense of being faithful to the model. The latter activity is called *verification* and has nothing to do with the real world, at least not directly.

Models are either *stochastic* or *deterministic*. Intuitively, a stochastic model assumes uncertain or probabilistic inputs regarding an experiment or a situation and makes an indefinite prediction of the results. A deterministic model states exactly what will happen, as if there were no uncertainty. More formally, a stochastic model requires the terminology of the theory of probability (events, random variables and probabilities, see Appendix A) for its description, whereas a deterministic model does not. The contrast is particularly important for Lanchester models (Chapter 5), where the same data can be employed by both types of model. Most combat models considered in this book are stochastic.

The DoD glossary points out that the terms "modeling" and "simulation" are often used interchangeably, but nonetheless offers a separate definition of simulation. Our habit in this book will be to use the word *simulation* only to refer to a particular method for implementing stochastic models that generates a sequence of replications of an abstract experiment, making inferences from the sequence. The most important kind of simulation is *Monte Carlo simulation*, as described in Appendix C. This type of model is one of seven shown in Fig. 1. The three types to the right of it are also simulations, by our definition, since all face the problem of generating sufficient replications in an uncertain environment to enable reliable inference. The models differ in the way in which replications are generated and managed, whether computationally or manually. Contrary to simulations, *analytic models* do not rely on multiple replications, instead developing formulas that describe results in a concise way. Analytic models may or may not be stochastic.

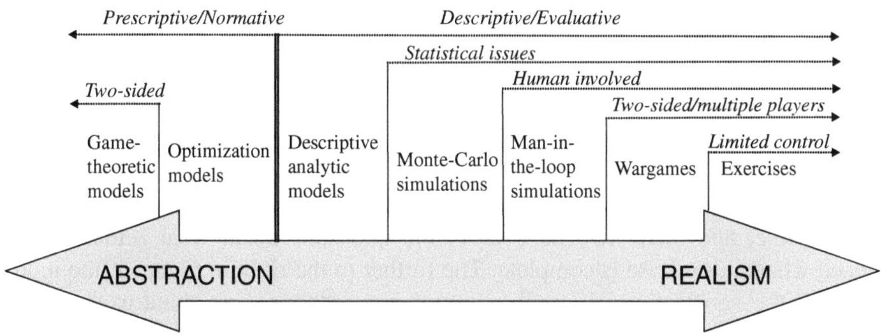

Figure 1: The combat modeling spectrum

Figure 1 shows the "combat modeling spectrum" with an emphasis on the trade-off between abstraction and realism. The primary division is between models that explicitly aim to find optimal decisions (*prescriptive* or *normative* models on the left side) and models that do not (*descriptive* or *evaluative* models on the right side). Prescriptive models aspire to prescribe optimal courses of action in complex combat situations. It is natural to expect that such models will require greater abstraction in order to find a useful and tractable implementation. Descriptive models describe combat phenomena and processes without prescribing courses of action, so they can be less abstract. The objective of a descriptive model is to gain insights about the main components of a system or the evolution of a combat situation, analyzing cause-and-effect relations and comparing alternative courses of action. Sometimes, in rare situations where the uncertainty is limited and well defined, and the model is supported with sufficient reliable data, the descriptive models may also be used for predictions of combat outcomes. Prescriptive models are analytic models with special ambitions. *Optimization* models deal with only a single decision maker, seeking a decision that is in some sense "optimal" (see Appendix B), whereas *game theoretic models* deal with multiple decision makers. *Wargaming* is a type of simulation that also involves multiple decision makers, using humans in an effort to achieve realism. Game theory and wargaming are introduced and contrasted in Chapter 7. At this point note only that these two methods for dealing with multiple decision makers are on opposite ends of the spectrum of abstraction.

Among descriptive models, *analytic* models, even when stochastic, are innately verifiable in that separate implementations should result in identical results. If two such implementations differ, then at least one of them is wrong. Simulations (the four types to the right in Figure 1) are generally formulated because an analytic model is either not available or intractable. If a stochastic model involves the use of a random number generator, then the model is either a Monte Carlo simulation or something further to the right. Because of the randomly generated replications, two correct Monte Carlo implementations can achieve different results, so verification is a more difficult subject than it is with analytic models (statistical issues are involved, as noted above the arrow in Figure 1). If a model involves even a single decision made by a flesh-and-blood human, then the implementation is either a *man-in-the-loop simulation* or something further to the right. If multiple humans with different goals make decisions, then the implementation is either a *wargame* or something further to the right. Finally, if any part of the implementation is directly affected by real-world phenomena, then the implementation is an *exercise*. Since exercises deal at least partially with the real world, some of their aspects are not controllable, and there may be irresolvable questions about what actually happened when an exercise is complete. The further to the right in Figure 1, the more difficult the verification. We will have nothing further to say about exercises in this book, and very little to say about man-in-the-loop simulations.

A given abstraction of a certain real-world situation can sometimes be implemented in multiple ways. Consider the real-world problem of flipping a fair coin three times, noting whether the event "two or more heads" happens. An analytical model would use the theory of probability (e.g., the Binomial distribution, see Appendix A) to conclude that the probability of the event is exactly 0.5. A Monte Carlo model of this situation is also possible where the three flips are simulated using random numbers, recording the fraction of the replications on which the event happens. The analytical model is surely best for this small problem, since the exact answer is derivable in a transparent manner with very little effort. In more complicated circumstances, the Monte Carlo model might be superior on the grounds of either efficiency or transparency. This particular question – whether an analytic or a Monte Carlo implementation should be pursued – is important and frequently occurring in studying stochastic models of combat. It will frequently be encountered in the sequel. The two methods are not incompatible – one useful verification tactic is to build both versions and compare them.

1.3 Modeling Shortcuts

As mentioned previously, most combat models considered in this book are stochastic. The inherent uncertainty of combat usually demands the type. Such models can be very complex, and therefore, in order to be able to implement them, sometimes simplifying "shortcut" assumptions are imposed. While these assumptions are powerful and facilitate efficient model implementation, they are also dangerous if not controlled and interpreted correctly. In this section we review some of the more commonly used shortcuts.

1.3.1 Expected Value Analysis

Many combat models are multi-stage in the sense that stage 1 determines some important input (random variable X, say) to stage 2. X might be the number of attackers that survive an initial defense or the number of helicopters available for the second stage of a mission after surviving a sandstorm in the first stage. Let $Y = f(X)$ be some scalar measure of the overall success of both stages and imagine that X is random. Since X is random so is Y. We would like to know the expected successfulness of the two-stage operation or $E(Y)$ (the expected value operator E () is introduced in Appendix A). The trouble is that finding $E(Y)$ requires us to evaluate $f(X)$ for all possibilities of X, and there may be many of them. If the function $f(X)$ is difficult or expensive to evaluate, it will be tempting to evaluate only $f(E(X))$, the overall successfulness of the second stage given the average output of the first stage, instead of the more complicated $E(f(X))$, the average successfulness of the two-stage operation. The systematic replacement of random variables by their expected values in this manner is called *Expected Value Analysis* (EVA),

EVA is often harmless. We are told that "someone dies on a highway in the United States every 13 min." That statement is not true. It would take a tremendous and pointless coordination effort to make it be true that highway deaths follow each other by exactly 13 min. What *is* true is that the *average* interval between highway deaths is 13 min. The statement simply omits the word "average", which is harmless as long as it is only used to estimate highway deaths over long intervals of time. However, EVA is not always harmless. Consider a system of n redundant controls designed to make sure that a nuclear attack does not start accidentally. Each control works independently with probability 0.9, and the system works as long as at least one control works. By EVA, we could argue that two such controls suffice. If X is the number of functional controls, then $E(X) = 2(0.9) = 1.8$, which is safe because 1.8 exceeds 1.0. There are two things wrong with this EVA analysis, one cosmetic and one fundamental. The cosmetic problem is that clearly the number of functional controls cannot be 1.8, since it is always an integer. The fundamental problem in this example is that the EVA outcome is insufficient to tell us the probability of an accidental nuclear attack, which is far too large when only two controls are employed (0.01, to be exact). As another example consider a tank crossing a minefield at night. There is a clear passage through the minefield but the disoriented driver of the tank may drive either 10 m right or left of the clear passage, with probability 0.5 either way. Using EVA, we might conclude that the tank always drives exactly on the clear passage, since the average deviation is 0, and is therefore safe, while in reality it *never* drives on the clear passage and could very well get blown up. In addition to losing all track of variability as shown in the first example, the deterministic "equivalent" model achieved by EVA can be wrong even on the average, as shown in the second.

If EVA is such a bad idea, then why do people do it? The primary answer is that EVA leads to deterministic models, and deterministic models are simpler than stochastic models, both conceptually and computationally. The point is sometimes made that combat models are inevitably only poor approximations to reality, so much so that approximating certain random variables by their average values is not worth worrying about. That is a good point, but there are some strong counter-arguments to it (Lucas, 2000). The introduction of EVA can turn an approximate but useful model into one that is actually misleading.

Our point here is not that EVA is something always to be avoided, but only that the analyst should be aware of the dangers. The possible employment of EVA is one of the significant decisions that an analyst must make in formulating a stochastic model, so it is important to at least have a name for it. This subject is taken up again in Chapter 5, where the stochastic and deterministic versions of Lanchester's equations are compared.

1.3.2 Universal Independence

Stochastic models generally deal with the interactions of a variety of different phenomena, and the statistical relationships between them are important to both model verity and model tractability. It is nearly always true that the most tractable assumption is that the phenomena are all independent of each other, so much so that the habit of making this assumption deserves a name: *Universal Independence* (UI). There are two basic reasons for UI's popularity:

(1) It often is not a bad assumption about the real world. There are lots of phenomena that simply have nothing to do with one another in any practical sense, in spite of fables about the outcome of a battle depending on the flight of a butterfly thousands of miles away.

(2) The theory of probability is usually at its simplest when dealing with independent events and random variables. The variance of a sum of random variables, for example, is equal to the sum of the individual variances when the random variables are independent, but not otherwise.

In spite of the first reason, there are circumstances where the assumption of independence is dangerously bad. For example, let E and F be the events that tanks 1 and 2 survive a transit of a minefield. Are E and F independent? They might be if the tanks travel independently, but they are certainly not independent if tank 2 simply follows the path of tank 1. If tank 1 survives, then its path must be safe, so tank 2 can also survive by following the same path. The validity of the independence assumption thus depends on how tanks actually travel through minefields. Since one of the principal countermeasures to a minefield is for those forced to cross it to all follow the same path, a model based on UI is likely to be highly optimistic about results from the viewpoint of the minefield planner. This observation has been the cause of considerable turmoil in software for minefield planning (see Chapter 8 for details). Another bad application of UI would be when several shots from a certain weapon share the problem that the weapon's fire control system is biased, resulting in a common aiming error (see Chapter 2).

Our point here is not that the assumption of independence is a bad idea; indeed, we will frequently employ it in the models developed in this book. It is instead the *casual* assumption of independence, as if the assumption were somehow a standard that ought not to be disputed, that we wish to warn of. The popularity of UI stems from the tractability that it almost always lends to models of combat. Validity, however, is something that ought to be carefully considered before employing it.

1.3.3 Tuning Parameters

Combat models sometimes incorporate parameters whose exact meaning is hard to divine, even with an English description. Possible descriptions might be "propensity to communicate," "shrink factor," or "surrender coefficient." Such parameters are typically present because combat models must sometimes deal with aspects of the real world that cannot be ignored, even though they are poorly understood. These parameters are sometimes adjusted by experts so that model results are more or less valid; i.e., "tuned." The problem is that model results can sometimes be quite sensitive to these parameters, which is dangerous because they represent poorly understood phenomena. The worst situation is when the model is being used to advocate some new tactic or combat system, since tuning parameters can be adjusted to make novelty appear advantageous.

Extreme assumptions are often easier to model than intermediate assumptions, so a tuning parameter might be a bridge from one extreme to another that effects a realistic intermediate compromise. An example of this occurs in trying to predict the probability that an array of several sensors will detect a target, given detection probabilities for the individual sensors. We might apply UI and model all the detection events as independent, an assumption that is almost surely optimistic if the sensors are located physically near each other. An equally extreme assumption in the pessimistic direction would be that all sensors are completely dependent, in which case the whole array is equivalent to whichever single sensor has the largest detection probability. Having found an optimistic answer A and a pessimistic answer B, we might invent an "independence factor" f and let the array's detection probability be $A(1-f)+Bf$. As f moves from 0 to 1, the detection probability moves from optimistic to pessimistic. Ideally, f would be determined in experiments involving actual sensor arrays and actual targets or at least tuned by an expert so that the predicted detection probabilities seem realistic. The tuning factor might temporarily achieve realism, but it also disguises our ignorance of how detections actually happen. Suppose someone invents a new sensor array whose elements are much closer together than the array for which the model was tuned.

We ought to be suspicious that the tuning factor for the new array should be smaller, but have no means of judging how much smaller, and might very well leave f unchanged because it is too much trouble to worry about.

Necessity often dictates the use of tuning factors, but the reader should be alert to the dangers. Whenever you discover something called the "communications adjustment factor" in a combat model, a skeptical investigation is appropriate.

1.3.4 The Ostrich Effect

Another alternative when encountering a poorly understood or controversial phenomenon in the course of formulating a combat model is to simply ignore it. Since the phenomenon is not even addressed, this tactic may forestall some of the skeptical questions likely to arise when using a tuning factor. Ignoring the phenomenon, and thus not assigning any values to the parameters that represent it, may very well generate less controversy than assigning them questionable values.

An example of this effect is presented in Morse and Kimball (1950). At one time merchant ships traveling across the Atlantic were given torpedo nets to entangle torpedoes launched by U-boats. The nets worked, but the analysts of the time concluded that they were a bad idea because they cost more than the ships and cargoes that they saved. In coming to this conclusion, the analysts ignored the fact that the nets also saved the lives of merchant seamen, effectively valuing those lives at zero. The analysts' recommendations were taken (the torpedo net idea was abandoned), but imagine what might have happened if the lives had been valued at (say) $100 each. Any value on human life is controversial, so the whole process might have become hung up on debates about whether $100 is the right number. Surely $100 is closer to the value of a human life than $0, as the analysts effectively assumed, but assuming $0 forestalled the debate by not even bringing up the subject.

It is sometimes not obvious whether a certain "detail" can be ignored or not. Suppose you are investigating how a diesel-electric submarine should evade a pursuer while submerged. Such submarines are dependent entirely on the energy stored in their batteries, so it might be tempting to model the batteries as an energy supply E that can be expended at whatever rate the submarine chooses, as long as energy consumed never exceeds E. The problem is that the submarine's best escape tactic might very well involve high speeds, therefore high power consumption, and therefore high battery currents. Batteries can waste a significant fraction of their power on internal resistance whenever the withdrawn current is high, which leads to less energy available for avoidance. You therefore have two choices:

- Persist with the original assumption, ignoring the effects of battery resistance.
- Find out more about batteries.

The first choice corresponds to employing the ostrich effect to internal resistance, since you will never have to mention the subject. The second is likely to require a large investment of time, since it turns out that there is much more to learn about batteries, some of which is likely to be classified. You have reached a crucial point in the process of constructing a model of the situation.

A model, being an abstraction, will always involve ignoring certain aspects of the real world. Many of these omissions are obviously harmless, but some are not. It is important to remember what has been omitted, and perhaps to test for significance after analysis is complete. In the battery example above, it may turn out that the submarine does not want to go fast for other reasons, even if its battery has no internal resistance, so the assumption that there is no resistance is harmless. It could also turn out that the details of how batteries work are essential if one wants to advise submarines on how to evade a pursuer. The latter possibility should be borne in mind, since the significance of an abstraction may not be clear until the model is used for something. The ostrich effect sometimes makes us forget that there ever was an issue, and that is the danger.

1.3.5 Convenient Distributions

There are a few probability distributions that are commonly used in combat models because of their tractability and the elegant results they can produce. Among these distributions we find the *Poisson*, *exponential*, *uniform*, and *normal* distributions (see Appendix A). Behind each such distribution hides several assumptions that may or may not be applicable to the system or situation being modeled. For example, the memoryless property of the exponential distribution is often analytically convenient, to the point where it is sometimes assumed without justification. This habit is potentially dangerous. In some situations it is a reasonable approximation to the actual behavior of a system, but in others (e.g., the "up" time of a machine that is subject to a mechanical wear) this may not be the case.

Another example is the normal distribution that is widely used in firing theory (see Chapter 2). There is strong empirical evidence that justifies the use of this distribution in certain firing situations (e.g., artillery); however, in other firing situations (e.g., precision-guided missiles) this may not be the case. As with the other shortcuts mentioned above, caution must be exercised to make sure that the choice of probability distribution is appropriate for the situation.

1.4 Notation and Conventions

The following notation is used throughout the book.

$\sum$ means sum, $\prod$ means product, $\int$ means integral; and $\frac{d}{dx}$ denotes a derivative with respect to x. Thus,

$$\sum_{i=1}^{4} i = 10, \quad \prod_{i=1}^{4} i = 24, \quad \int_{0}^{4} x dx = 8, \text{ and } \frac{d}{dx}(x^2/2) = x.$$

When no limits are specified and context makes clear what "everything" means, $\sum$, $\prod$, and $\int$ are "over everything."

When no limiting statement is made, "for all" should be understood as "for all possible values".

max {2, 1, 7, 3} is 7, and more generally $\max_{s \in S}(a_s)$ is the largest value of a_s among all those with subscript s in the set S. Similar notation is used for the min operator.

Scalars are presented in *italic*, vectors and matrices are **bold**. Thus $\mathbf{x} = (x_1, x_2, x_3)$ is a vector with three components. If the number of components is not important, a vector might also be abbreviated $\mathbf{x} = (x_i)$.

$E(X)$ is the expected value of random variable X, and $P(A)$ is the probability of the event A. For more detailed probabilistic notation, see Appendix A.

Section 7.3.2 is the second subsection of the third section of Chapter 7, etc.

Figures, tables, examples, and exercises are numbered sequentially in each chapter.

References are accumulated at the end of the book, with referring sections or subsections given in [square brackets].

1.5 Book Overview

The rest of this book comprises nine chapters and three appendices, each of which is previewed below.

Chapter 2: Shooting without Feedback

Fire is the main manifestation of military force and the major cause for combat attrition. In this chapter we address the question of what happens when a weapon delivers rounds of fire on a target but gets no feedback regarding the effect. Issues of accuracy and damage are addressed in the cases of single and multiple shots, for single and multiple targets. While the majority of the chapter covers descriptive models, optimal shooting tactics are discussed too.

Chapter 3: Shooting with Feedback

This chapter deals with the situation where the shooter gets feedback about the status of the target (killed, partially damaged, unharmed) and possibly the miss distance. This feedback may be subject to error. In the presence of feedback, the question of optimal shoot-look-shoot (SLS) tactics arises. Both descriptive and prescriptive SLS models are presented.

Chapter 4: Target Defense

Consider a group of attackers approaching a target, hoping to kill it by overwhelming its defenses. The defense is armed with some anti-attacker weapons called interceptors, each of which can kill only the attacker to which it is assigned. The goal is to use the interceptors to maximize the survival probability of the target. Models, both descriptive and prescriptive, for the defense of single and multiple targets are presented.

Chapter 5: Attrition Models

The purpose of firing at the enemy is to cause attrition to his forces. In this chapter we present aggregate attrition models that apply to force-on-force situations. Other important ingredients in combat such as morale, endurance, maneuver, logistics, command and control, and intelligence are ignored or mentioned only in passing. The chapter focuses on *Lanchester Models,* both deterministic and stochastic, which are widely used in force-on-force modeling.

Chapter 6: Game Theory and Wargames

Another major factor that affects the outcome of combat situations is the enemy's behavior. The uncertainty associated with this behavior is different from the uncertainty that is associated, say, with the trajectory of a projectile. The projectile has no "feelings" or "opinions" about the combat outcome and its behavior is determined by Mother Nature, but the enemy has objectives and strong feelings about the outcome and behaves rationally. Thus, the uncertainty about enemy actions deserves a different kind of treatment; decisions made against an enemy must differ qualitatively from decision made against Mother Nature. This chapter presents two types of models that address the case of a decision-making enemy: *game theoretic* models prescribe optimal strategies for both sides, while *wargames* are human-in-the-loop descriptive models (See Figure 1).

Chapter 7: Search

This chapter is about searching for physical objects with sensors that act very much like our eyes and ears. Similarly to our senses, the closer the sensor is to the object the higher is the probability of detecting it. Roughly speaking, search is a sequence of repeated trials where successes happen when the circumstances (e.g., distance, visibility, effectiveness of sensor) are such that the target is detected. This chapter reviews descriptive and prescriptive models for detecting the target as soon as possible. The models apply to both stationary and moving targets.

Chapter 8: Mine Warfare

A mine is a distinctive type of weapon; it is stationary once laid, its operation is triggered by the target, rather than the attacker, and it destroys itself in the process of attacking its target. Mines rely for their effectiveness on the enemy's need to move, and they are most effective when operating as a group – a *minefield.* In this chapter we introduce simple minefield models and then proceed to discuss minefield planning and minefield clearing. Game theoretic models are considered.

Chapter 9: Unmanned Aerial Vehicles

An unmanned aerial vehicle (UAV) is a remotely piloted or self-piloted aircraft that can carry a variety of payloads such as sensors, communication equipment, and even weapons. UAVs are increasingly becoming key weapon systems in modern warfare. Typical missions of UAVs are surveillance, reconnaissance, target engagement and fire control for other long-range weapons, and attack. There are several problems associated with the design (size, velocity, payload, flying altitude, sensing capabilities) and operation of UAVs. This chapter only addresses operational issues. Specifically, two types of UAV problems are considered: (1) routing and scheduling UAVs in reconnaissance and search missions, (2) evaluating the effectiveness of armed UAVs in attack missions. For the former, several optimization models are developed and demonstrated. For the latter, descriptive stochastic models are presented for evaluating the effect of such weapons in various engagement scenarios.

Chapter 10: Terror and Insurgency

Terrorism and insurgency are profoundly different from the combat situations discussed thus far. First, they involve non-state actors that are organized and equipped differently than conventional forces and are not bound by any combat rules. Second, the general population, which only plays a passive role in conventional warfare, is much more involved in terrorist and insurgency actions both as victims (of terrorist acts) and active supporters or deniers (of an insurgency). This chapter presents three descriptive models. The first model evaluates the effect of suicide bombing in a crowded area. The second model describes the dynamics of a biological attack and the effect of various response strategies. The third model presents the effect of intelligence in counterinsurgency operations.

Appendices

Appendix A reviews basic probability concepts, probability distributions and Markov models. Appendix B introduces optimization models, and Appendix C discusses Monte Carlo simulations.

Chapter 2
Shooting Without Feedback

One or two sarcastic spirits
Pointed out to him, however,
That it might be much more useful
If he sometimes hit the target.

From "Hiawatha Designs an Experiment", a statistical poem by Maurice Kendall (1959)

2.1 Introduction

In combat, man specializes in exerting lethal force at a distance. The mechanism for exerting this force has progressed from stones to spears to firearms to rockets, but the basic shooting problem for the marksman has always been to effectively combine accuracy and lethality at long range. This chapter is devoted to abstract models of the shooting process. Such models have several possible purposes. One purpose is simply to determine the probability of killing the target for use in a higher level combat model. Another is to influence the design of a weapon system – this purpose will be aided by the dependence of kill probability on fundamental parameters that quantify accuracy and lethality. A third purpose is to influence the shooting process itself, since there are tactical decisions to be made about where to aim and how often to shoot. We assume throughout that the marksman gets no feedback between shots, so questions about how to adjust the aim point or when to stop shooting must be reserved until Chapter 3.

The objective of the shooting process is to kill the target. The event "kill" may have several meanings such as "immobilize," "damage," "find," or even "rescue," but the only important thing is that the event should either happen to the target or not with each shot, with no residual effect if the target is not killed.

Each shot is aimed at a certain aim point that the marksman controls, but the shot usually hits a different impact point because of dispersion errors that are assumed to be independent from shot to shot. The marksman may have a faulty notion of where the target is located, with the difference between the marksman's estimate and the actual target location being the target location error, one of several possible contributors to a bias error that affects all shots. We assume throughout that dispersion errors and bias errors are normally distributed in two dimensions. The world that we know has three dimensions, but most targets are known to lie on the surface of the Earth somewhere, which fixes the vertical dimension. Even problems that involve aerial or submarine targets have in most cases been reduced to two-dimensional problems by the invention of the proximity fuse.

A. Washburn, M. Kress, *Combat Modeling*, International Series in Operations Research & Management Science 134, DOI 10.1007/978-1-4419-0790-5_2,

While bivariate normality is not always the case in reality, it is usually close to the truth, except for outliers, and anyway the normal distribution should still be our default assumption because of the central limit theorem (see Appendix A).

The plan of this chapter is to first deal with problems involving only a single shot, and then generalize to multi-shot problems. Section 2.2 deals with problems where there is only one shot. Section 2.3 generalizes to include many shots, as long as they are all independent. Section 2.4 deals with multi-shot problems where bias errors are significant and thus the shots are not independent. In such situations pattern firing is useful.

It will be useful to have the *Chapter2.xls* Excel™ workbook available.

2.2 Single-Shot Kill Probability

In this section we present some basic firing models that describe what happens when a single shot is fired. The shot will kill the target if, in spite of various errors in firing and in locating the target, the shot's distance from the target is sufficiently small. Our object is to determine the probability of kill as a function of more basic parameters.

2.2.1 Damage Functions and Lethal Area

Definition: The *miss distance* r is the distance between the impact point of the shot and the location of the target.

Definition: The *damage function* $D(r)$ is the probability that the target is killed by the shot, as a function of the miss distance r. "Kill function" might be a more natural name, but our usage reflects common practice. Figure 1 shows three examples of damage functions.

The damage function has only one argument, so we are implicitly assuming a radially symmetric situation where damage to the target is invariant to the position of the impact point, except for the miss distance. Damage functions are in practice measured through some combination of theory and experiment. In this book, we will take them to be given.

The damage function is a conditional kill probability. The unconditional kill probability P_K is obtained by averaging over the miss distance. Let $f(x, y)$ be the bivariate density function of the position of the weapon's point of impact, relative to the target's location. Then, since $r = \sqrt{x^2 + y^2}$,

$$P_K = \iint D\left(\sqrt{x^2 + y^2}\right) f(x, y)\, dx\, dy, \tag{2.1}$$

where the lack of limits means that the integral is to be taken over the whole plane. Sections 2.2.2 through 2.2.4 deal with various special cases of (2.1).

If the position of the target were uniformly distributed within some large area A, then (2.1) would be (substituting $f(x, y) = 1/A$),

$$P_K = \frac{1}{A} \iint_A D\left(\sqrt{x^2 + y^2}\right)\, dx\, dy\,, \tag{2.2}$$

where the notation indicates that the integral is now taken only over the area A. However, since A is by assumption large, (2.2) is approximately the same as $P_K = a / A$, where

$$a = \iint D\left(\sqrt{x^2 + y^2}\right)\, dx\, dy \quad \text{or} \tag{2.3}$$

$$a = 2\pi \int_0^\infty r D(r) dr\,. \tag{2.4}$$

Formula (2.4) was obtained from (2.3) by introducing polar coordinates. The quantity "a" is the *lethal area* of the weapon and serves as a scalar measure of weapon effectiveness with respect to a certain target.

Although it is not logically necessary, the damage function $D(r)$ is typically a non-increasing function of its argument. As long as this is true, it is sometimes convenient to describe a damage function as follows: imagine that the weapon has a random "lethal radius" R associated with it, and that a target will be killed if and only if it lies within a distance R of the weapon's point of impact. Recalling the meaning of $D(r)$, it must evidently be the case that

$$D(r) = P(R > r). \tag{2.5}$$

If $D(r)$ is differentiable, one can go further and discover the probability density function of R:

$$f_R(r) = -\frac{d}{dr} D(r)\,. \tag{2.6}$$

The area covered by the weapon is πR^2, so it should come as no surprise that $a = \pi E(R^2)$, where $E(\ )$ denotes expectation. This provides another interpretation of the term "lethal area."

2.2.2 Cookie-Cutter Damage Function

The conceptually simplest kind of weapon is one for which the lethal radius R is a constant, in which case the lethal area is $a = \pi R^2$. If the firing errors are circular normal (by which we mean that the standard deviation of the error in all directions is the same number σ) and centered on the target, then the two-dimensional density function of the error is $\frac{1}{2\pi\sigma^2}\exp(-\frac{x^2+y^2}{2\sigma^2})$, and (2.1) reduces after the introduction of polar coordinates to

$$P_K = \int_0^{2\pi}\int_0^R \frac{1}{2\pi\sigma^2}\exp(-\frac{r^2}{2\sigma^2})r\,dr\,d\theta = 1-\exp\left(-\frac{1}{2}R^2/\sigma^2\right). \tag{2.7}$$

This is a very simple expression for kill probability. A large number of military analyses have been based on it. Since (2.7) is a formula for the probability that the miss distance does not exceed R, it is also a formula for the cumulative distribution function of the miss distance when firing errors are circular normal, a Rayleigh type of random variable.

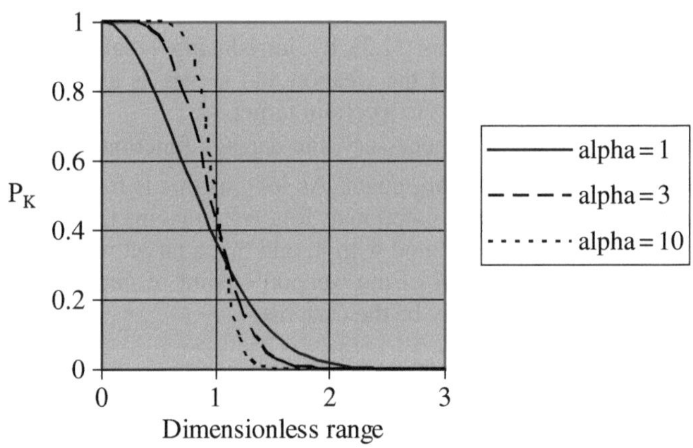

Figure 1: The Diffuse Gaussian (alpha=1) damage function is contrasted with two more definite damage functions with higher values of alpha. All three have the same lethal area. Dimensionless range is the ratio of miss distance to the scale parameter b.

Unfortunately, most departures from the circular normal assumption about errors result in significantly more complicated expressions for P_K. If the circular normal error distribution is offset from the target by some distance h, for example, then evaluation of (2.1) involves the integral of a Bessel function that must be done numerically. One of the early products of the RAND corporation was a table of such probabilities (Marcum, 1950). Sheet "OffsetQ" of *Chapter2.xls* employs the function $OffsetQ(R/\sigma, h/\sigma)$ to calculate the miss probability (the probability that the miss distance exceeds R) in this case. If the normal distribution is centered on the target, but not circular, then the function $EllipQ(R, \sigma_1, \sigma_2)$ can be used instead. VBA code for both functions can be found in module 1 of *Chapter2.xls*. The code is based on Gilliland (1962), which also covers the general case where the normal distribution is neither centered nor circular.

Example 1: If $R = 100$ m and $\sigma = 50$ m, the kill probability according to (2.7) is 0.865. If the aim point is offset from the center of the distribution by $h = 25$ m, then the miss probability is $OffsetQ(2, 0.5) = 0.169$, so the kill probability is only 0.831. The bigger the offset, the smaller the kill probability. See sheet "OffsetQ" in *Chapter2.xls* for more calculations of this kind.

Example 2: Suppose $R = 100$ m, but that the standard deviation of the error along the line between marksman and target (the downrange direction) is 90 m, while the standard deviation of the error in the direction perpendicular to that (the crossrange or deflection direction) is only 30 m. It is typical for downrange and crossrange errors to differ by about a factor of 3, as assumed here. The miss probability is $EllipQ(100, 90, 30) = EllipQ(100, 30, 90) = 0.293$, so $P_K = 0.707$. An approximation can be made by letting the error in both directions be whatever is required to preserve the product of $(30\text{ m})(90\text{ m}) = (2700\text{ m}^2)$. That error is $\sigma = 51.96$ m. Employing (2.7), we find that $P_K = 0.843$ and see that it can be significantly optimistic to assume that errors are circular when they are not. Sheet "EllipQ" of *Chapter2.xls* generalizes this example.

Formula (2.7) is sometimes expressed in the form

$$P_K = 1 - (0.5)^{\left(R^2/\text{CEP}^2\right)}, \tag{2.8}$$

where CEP stands for "circular error probable."

Definition: *Circular Error Probable* (CEP) is the radius of the smallest circle that contains the two-dimensional error with probability 0.5.

In (2.8), CEP references a circular normal firing error, but the definition of CEP applies to any kind of a two-dimensional error distribution. For a circular normal distribution, CEP is related to σ by CEP = $\sigma\sqrt{2\ln 2}$ = 1.1774σ (see Exercise 14).

Dealing with CEP instead of σ has the advantage of easing explanations to novices, since the idea of standard deviation is not involved. CEP can itself be confusing, however. The corresponding notion in one dimension would be an interval that contains the error half the time, the half-length of which is sometimes called the linear error probable or LEP. A novice might expect LEP and CEP to be equal, since each corresponds to a region that contains half the shots, but they are not. A square with side 2 × LEP will contain only $(0.5)^2$ of the shots, since both coordinates must fall within the relevant side of the square. A circle with radius LEP, since it falls within that square, will contain even less than 25% of the shots. CEP is thus considerably greater than LEP, and SEP (spherical error probable) is larger yet. In other words, the magnitude of the "error probable" must depend on the number of dimensions that one is interested in. It is simpler to say that the error has standard deviation σ along any line, regardless of the number of dimensions.

2.2.3 Diffuse Gaussian (DG) Damage Function

The DG damage function has the form $D(r) = \exp\left(-\frac{r^2}{2b^2}\right)$ for some scale factor b. Unlike the cookie-cutter damage function, there is a positive probability of killing the target no matter what the miss distance. The lethal area of such a weapon is $2\pi b^2$. Figure 1 compares $D(r)$ for a DG damage function with two other damage functions that will be discussed in the next section. The weapon with the DG function is evidently "sloppier" than the others, in the sense that results at any given range are harder to predict. Whether this feature makes the DG assumption more realistic depends on the damage mechanism. Weapons that kill by fragmentation (e.g. artillery using "fragmentation" rounds) generally have a sloppier damage function than those that kill by overpressure (e.g., artillery using high explosive rounds). The cookie-cutter damage function is the least sloppy of them all because it suddenly falls from 1 to 0 as the miss distance increases.

The DG assumption combines very nicely with the assumption of normal errors to produce a simple, general expression for P_K. If the center of the error distribution is (μ_X, μ_Y), and if the standard deviations of the X and Y errors are (σ_X, σ_Y), then (2.1) can be evaluated analytically. In fact, (2.1) can be integrated in closed form even when the damage function is not radially symmetric, so we record here the result for the asymmetric DG damage function: If $D(x, y) = \exp\left(-\frac{1}{2}\left(\left(\frac{x}{b_X}\right)^2 + \left(\frac{y}{b_Y}\right)^2\right)\right)$, then

$$P_K = \frac{b_X b_Y}{\sqrt{(b_X^2 + \sigma_X^2)(b_Y^2 + \sigma_Y^2)}} \exp\left(-\frac{1}{2}\left(\frac{\mu_X^2}{b_X^2 + \sigma_X^2} + \frac{\mu_Y^2}{b_Y^2 + \sigma_Y^2}\right)\right). \quad (2.9)$$

In the special case where $\mu_X = \mu_Y = 0$, $b_X = b_Y = b$, and $\sigma_X = \sigma_Y = \sigma$, (2.9) reduces to

$$P_K = \frac{b^2}{b^2 + \sigma^2}, \quad (2.10)$$

which is comparable to (2.7).

There is no cookie-cutter counterpart to (2.9); that is, there is no simple analytic expression for the cookie-cutter kill probability with the generality of (2.9). While it is true that the cookie-cutter damage function is conceptually simpler than the DG damage function, it is equally true that the DG function is analytically simpler than the cookie-cutter function. The simplicity of the DG assumption was first taken advantage of in 1941 by von Neumann (Taub, 1962), who used it in determining optimal bomb spacing in World War II.

The "DGGenrl" sheet of *Chapter2.xls* calculates P_K from (2.9), as well as performing a Monte Carlo simulation whose object is to estimate the same quantity. The agreement of the two answers is part of the process of verifying that each is correct.

2.2.4 Other Damage Functions

It was pointed out in Section 2.1 that any non-increasing damage function can be interpreted as the probability law for a random lethal radius R. The Diffuse Gaussian damage function, for example, has associated with it the density function $f_R(r) = \frac{r}{b^2}\exp\left(-\frac{r^2}{2b^2}\right)$, which is a Rayleigh density function. It is perhaps more natural to deal with the random variable R^2, since R^2 is directly related to area covered. For the DG damage function, R^2 is an exponential random variable with mean $2b^2$.

It is possible, of course, to reverse the process: begin with some convenient density for R or R^2 and then discover the associated damage function by integration. One convenient class of damage functions (the Gamma class) can be obtained by assuming that $\frac{1}{2}R^2/b^2$ has the Gamma density $\frac{\alpha(\alpha x)^{\alpha-1}}{\Gamma(\alpha)}\exp(-\alpha x)$ for some $\alpha > 0$, in which case the DG damage function is the special case $\alpha = 1$ and the cookie-cutter damage function is obtained in the limit as $\alpha \to \infty$. Every member of the class has the same lethal area $2\pi b^2$. The associated damage function is

$$D_\alpha(r) = 1 - \Gamma\left(\alpha, \frac{\alpha r^2}{2b^2}\right) \tag{2.11}$$

where $\Gamma(\alpha, x)$ is the incomplete Gamma function (in Excel™, $\Gamma(\alpha, x)$ is GAMMADIST(x, α, 1, TRUE) — see Figure 1 or sheet "Gamma" of *Chapter2.xls* for plots of $D_\alpha(r)$ versus $\frac{r}{\sqrt{2}b}$).

The Gamma class is convenient because it has both scaling (b) and shaping (α) parameters, and also because there is a simple formula for P_K when the firing error is circular normal with standard deviation σ and centered on the target:

$$P_K = 1 - \left(1 + \frac{b^2}{\alpha\sigma^2}\right)^{-\alpha} \quad ; \quad \alpha > 0. \tag{2.12}$$

Formula (2.10) is the special case where $\alpha = 1$, and (2.7) is the limiting case as $\alpha \to \infty$. Except in the case $\alpha = 1$, where (2.9) applies, the centered circular normal assumptions are essential for having a simple expression for P_K.

Another class of density functions for R^2 with both a shaping and a scaling parameter is the class of lognormal densities. There turns out to be little to recommend this class in terms of analytic convenience – there are no counterparts to (2.11) and (2.12), for example. Nonetheless, the class has been widely used to model the effects of nuclear weapons (DIA, 1974).

2.3 Multiple-Shot Kill Probability

In this section we consider multiple shots made in a salvo.

Definition: Regardless of the number of targets, a "salvo" is a group of shots all taken on the basis of the same information, with no information feedback between shots. We often imagine that the shots are all made at the same time, even though that is not necessarily the case.

2.3.1 Simultaneous Independent Shots

Suppose that a salvo of n independent shots is fired at a target, and let q_i be the probability that the ith shot fails to kill it. The numbers q_i may be obtained from one of the formulas in Section 2.2 or by some other method. Since all shots are by assumption independent, the probability that all n miss the target is the product of the miss probabilities, so

$$P_K = 1 - q_1 q_2 \ldots q_n. \tag{2.13}$$

In the case where all the shots have the same miss probability q, this reduces to $P_K = 1 - q^n$, a formula that is used so often that it has a name "powering up." If each of three independent shots has a kill probability of 0.3, one might naively expect the kill probability of the collection to be 0.9. The correct answer, however, is $1–0.7^3 = 0.657$. The miss probability should be powered up, rather than making the kill probability proportional to the number of shots.

Formula (2.13) takes on a particularly simple form if the shots are all of the cookie-cutter type, and the firing errors are circular normal centered on the target. Let R_i and σ_i be the lethal radius and error standard deviation of the ith shot. Then $q_i = \exp\left(-\frac{1}{2}R_i^2/\sigma_i^2\right)$ from (2.7), and therefore

$$P_K = 1 - \exp(-X/2), \text{ where} \tag{2.14}$$

$$X = R_1^2/\sigma_1^2 + \cdots + R_n^2/\sigma_n^2 .$$

The quantity X can be thought of as a measure of the effectiveness of an arsenal of weapons against a particular target. The target dependence of this effectiveness can be eliminated if lethal radius scales in a known manner with the energy yield Y of the weapons. If the kill mechanism is overpressure, for example, then $R_i = KY_i^{1/3}$ where K is a target-dependent constant, and therefore $X = K^2\left[Y_1^{2/3}/\sigma_1^2 + \ldots + Y_n^{2/3}/\sigma_n^2\right]$. The quantity in [ ] is a target-independent measure of effectiveness for the group of weapons taken as a whole. It differs from "counter military potential" (CMP, see below) only in the scale factor required to convert standard deviation to circular error probable (CEP) for circular normal weapons (see Section 2.2.2).

The CMP of a group of n weapons is

$$\text{CMP} \equiv Y_1^{2/3}/\text{CEP}_1^2 + \cdots + Y_n^{2/3}/\text{CEP}_n^2 , \tag{2.15}$$

where yield is measured in equivalent kilotons of TNT, and CEP is measured in nautical miles. CMP is one of several measures that have been used to compare arsenals of nuclear weapons. This measure is very sensitive to accuracy – doubling all yields increases CMP by the factor $2^{2/3} = 1.6$, whereas halving all CEP's increases CMP by the larger factor $2^2 = 4$. In the 1970s, this fact was sometimes used to make the point that the relatively small but accurate nuclear arsenal of the United States was actually more potent than the large but inaccurate arsenal of the Soviet Union. Tsipis (1974), for example, estimated that CMP was 22000 for the US and 4000 for the SU in 1974.

An alternative measure of effectiveness for an arsenal is "equivalent megatons" (EMT). The definition of EMT is

$$\text{EMT} \equiv Y_1^{2/3} + \cdots + Y_n^{2/3} . \tag{2.16}$$

Since $Y_i^{2/3}$ is proportional to R_i^2, EMT is essentially a target-independent measure of the total lethal area of the arsenal. The "EMTCMP" sheet of *Chapter2.xls* compares the 1978 ICBM arsenals of the USA and the USSR using both measures.

If a total of C units of CMP are applied to a target, then the kill probability is of course still a function of the hardness of the target. For nuclear weapons making overpressure kills, with hardness h being measured in pounds per square inch (psi), an approximate formula (Matlin, 1972) valid for $30 \le h \le 1000$ psi is

$$P_K = 1 - \exp(-\,0.0583Ch^{-0.7}). \tag{2.17}$$

For example, a 1000 kiloton weapon with a CEP of 0.25 nautical miles will kill a 1000 psi target with probability 1 – exp(– (0.0583)(400)(0.00794)) = 0.52. Sixteen such weapons would be equally effective if the CEP were 1 nautical mile.

Figure 2 shows an application of the EMT idea to various historical ships, except that EMT is replaced by E8RPM. E8RPM is the equivalent rate of applying 8″ rounds per minute, plotted against firing range for each of five ships. Since yield (Y) is proportional to weight, and weight to the cube of dimension, $Y^{2/3}$ is proportional to dimension squared. Thus one 16″ round is roughly equivalent to four 8″ rounds, etc. Reducing all firepower to a common scale permits each ship to have a single "weight of broadside" versus range curve, independent of the target. Aircraft carriers can be portrayed on the same scale by calculating the rate at which aircraft can deliver 500-pound bombs, each of which is equivalent to two 8″ rounds. At a range of 20 nm, any of the battleships shown in Figure 2 could have quickly sunk the USS Enterprise. The battleship's problem in World War II was that an aircraft carrier could deliver enough firepower, even at long range, to put a battleship out of commission before any of its guns could come into play.

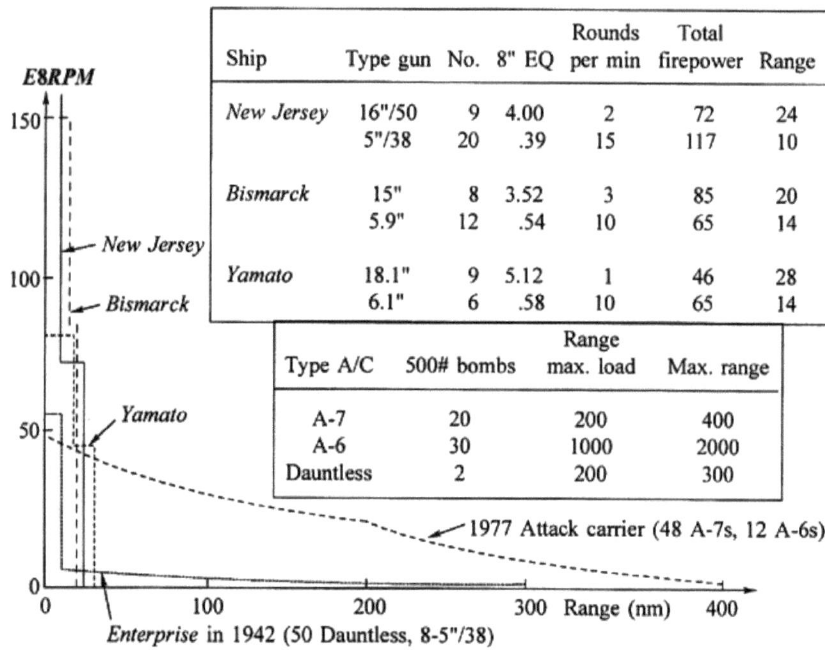

Ship	Type gun	No.	8" EQ	Rounds per min	Total firepower	Range
New Jersey	16"/50	9	4.00	2	72	24
	5"/38	20	.39	15	117	10
Bismarck	15"	8	3.52	3	85	20
	5.9"	12	.54	10	65	14
Yamato	18.1"	9	5.12	1	46	28
	6.1"	6	.58	10	65	14

Type A/C	500# bombs	Range max. load	Max. range
A-7	20	200	400
A-6	30	1000	2000
Dauntless	2	200	300

Figure 2: Equivalent 8″ rounds per minute for five historical ships.

Figure 2 does not tell the whole story, of course, since no information about accuracy or armor is given. Figure 2 does make it clear that the USS Enterprise and the USS New Jersey were very different weapon systems, and that the passage of 35 years produced an aircraft carrier much more powerful than the Enterprise.

2.3.2 Salvos of Dependent Shots

The firing errors dealt with in the previous section were *dispersion errors*, whereas in this section we assume the additional presence of a *bias error.*

Definition: *Dispersion errors* are firing errors that are independent and identically distributed among multiple shots. Such errors are sometimes also called ballistic errors.

Definition: *Bias errors* are errors common to all shots. Such errors are sometimes also called systematic errors or aiming errors.

Bias error might be due to a misalignment between the aiming and launching systems, to an error in target location, or to any other effect that introduces an error component common to all shots. The result is frequently that the impact points relative to the target are tightly grouped (indicating small dispersion errors) but in the wrong place due to bias, as in Figure 3. One can think of the bias error as being the center of gravity of the group, and of the dispersion errors as being deviations from the center of gravity. We shall use the notation that (σ_U, σ_V) are the (horizontal and vertical, say) standard deviations of the bias error, whereas the independent dispersion error for each shot has standard deviations (σ_X, σ_Y).

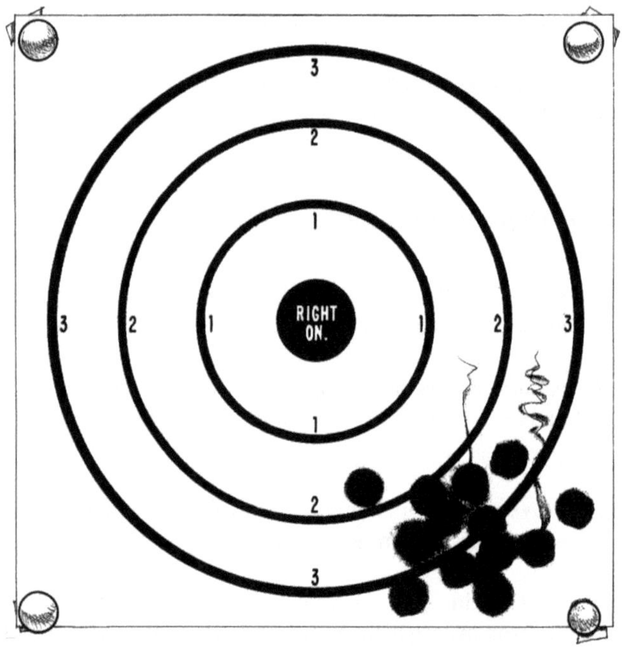

Figure 3: Illustrating a pattern with small dispersion and large bias. This marksman might be more effective if he had a larger (!) dispersion.

It is no longer possible to proceed by first finding the single-shot kill probability and then invoking (2.13) to obtain a simple expression for P_K, since the required independence assumption is falsified by the bias error. The following example should make this clear.

Example 3: Suppose there are two shots, and consider a one-dimensional problem where the bias error is X and the dispersion errors are Y_1 and Y_2 for shots 1 and 2. To keep the computations simple, assume that all errors are either –1, 0 or +1, rather than being normally distributed, and that the target will be killed if and only if the sum of the errors is 0 for some shot. Thus, the probability of killing the target with two shots is $P_K(2) = P(X + Y_1 = 0 \text{ or } X + Y_2 = 0)$. Assume that X is equally likely to be any of the three possibilities, whereas Y_1 and Y_2 are independent random variables with probabilities 1/6, 2/3, and 1/6 of being –1, 0, or +1. Using the theorem of total probability, the single-shot kill probability is $P_K(1) = P(X + Y_1 = 0) = (1/3)(1/6) + (1/3)(2/3) + (1/3)(1/6) = 1/3$. Each of the two shots will kill the target with probability 1/3. If we were to power up the miss probability, the kill probability for two shots would be $P_K(2) = 1 - (1 - P_K(1))^2 = 5/9$. However, powering up is not justified in this case because the two shots are not independent – they share the common error X. The correct answer can be obtained by conditioning on the value of X; that is, by powering up for each of the three possible values of X, and then averaging: $P_K(2) = (1/3)(1-(5/6)^2) + (1/3)(1-(1/3)^2) + (1/3)(1-(5/6)^2) = 1/2$. It is typical for the faulty computation to produce a kill probability that is too high.

In general, let (U,V) be the two-dimensional bias error, and let $f(u,v)$ be its density function. If there are n shots and $Q_i(u,v)$ is the miss probability of the ith shot, given that the bias error is (u,v), the expression for P_K is

$$P_K = E(1-\prod_{i=1}^{n} Q_i(U,V)) = \int\int (1-\prod_{i=1}^{n} Q_i(u,v)) f(u,v)\,dudv \tag{2.18}$$

We will have little direct application for (2.18), the main reason being the complicated nature of the functions $Q_i(u,v)$, which in general will depend on the aim points for each of the n shots. In fact, we will find no simple exact expressions for P_K in this section. The best that can be hoped for, other than solutions to specific problems that are important enough to justify the work involved in evaluating (2.18) for all aiming patterns of interest, is some rules of thumb that take the form of approximations. In deriving these approximations, it will be convenient to imagine that the only source of bias is an error in target location. The approximations are actually valid regardless of the source of bias or even if there are several sources (see Section 2.3.4).

Our first approximation to P_K is an upper bound obtained by making two unrealistic assumptions that are clearly favorable to the marksman. One assumption is that there are no dispersion errors, and the other assumption is that the marksman can exchange his weapons for any other weapon or weapons with the same total lethal area. The second assumption permits the marksman to avoid all of the problems associated with trying to fill up space with circles, since he can reshape the lethal area in any convenient manner. In the circularly symmetric case where $\sigma_U = \sigma_V = \sigma$, the marksman would prefer to have a single large cookie-cutter megaweapon that he would aim directly at the target or more precisely at the mean location of the target. If the total lethal area of n weapons is na, then the lethal radius of such a megaweapon would be $R = \sqrt{na/\pi}$, and the resulting kill probability would be (from (2.7)) $1-\exp\left(-R^2/2\sigma^2\right) = 1-\exp\left(-na/(2\pi\sigma^2)\right)$. More generally, the best megaweapon for our privileged marksman is a cookie-cutter with the same elliptical shape as the iso-probability contours of the error distribution. The resulting bound is

$$P_K \leq 1 - \exp(-z), \text{ where } z = \frac{na}{2\pi\sigma_U\sigma_V}. \tag{2.19}$$

Formula (2.19) was obtained by essentially assuming away all the overlap that is caused by dispersion errors, circle-packing problems, and noncookie-cutter weapons. The expression $1 - \exp(-z)$ should therefore be expected to be an accurate approximation in circumstances where overlap is expected to be a minor problem. Seven circles, for example, pack rather nicely into one circle without very much overlap. Figure 4, which is taken from the "Patterns" sheet of *Chapter2.xls*, shows the upper bound lying well above two other approximations that are introduced below.

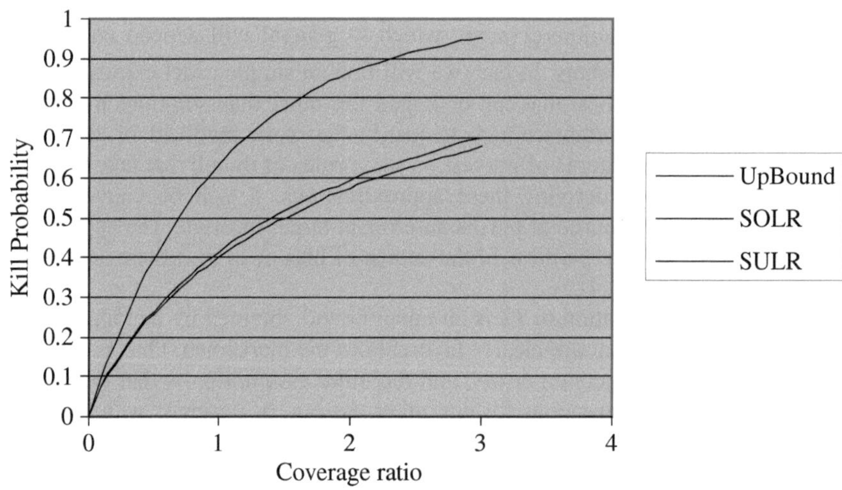

Figure 4: The two locally random approximations, SULR and SOLR, lie well below the upper bound on kill probability.

A different kind of approximation is based on the idea that overlap is inevitable, and that one should expect the amount of overlap to be whatever happens "at random." More precisely, the total lethal area *na* is assumed to be *in effect* so much confetti, with the marksman being able to control the density of confetti on a large "strategic" scale, but not the small-scale tendency of the flakes to overlap one another. Now, if *d* square inches of confetti are scattered uniformly at random on a one-inch square or in other words if the coverage ratio is *d*, then the fraction of the square that is covered is $1 - \exp(-d)$, as long as the flakes are sufficiently small. This expression is most easily derived by imagining that there are *k* independently located flakes of confetti with total area *d*. The probability that each flake covers the target is d/k, the fraction of the square covered by the flake. By powering up, the probability that some flake covers the target is $1-(1-d/k)^k$. The limit of this expression as k approaches infinity is $1-\exp(-d)$. This is true even if the density of confetti (d) depends on location. The marksman's problem is then to distribute a fixed amount of confetti (na) over the plane in such a manner that the (unconditional) kill probability is maximized.

Assume that $\sigma_U = \sigma_V = \sigma$, and that the marksman scatters all na units of confetti uniformly over a circle with radius r in the hope that some flake covers the target. This is the strategically uniform, locally random (SULR) case. Within the circle, the coverage ratio is $d = na/(\pi r^2)$. The probability that the target is actually in the circle is (from (2.7)) $1-\exp(-r^2/2\sigma^2)$, so the unconditional kill probability is

$$p(r) \equiv \left[1-\exp\left(-r^2/2\sigma^2\right)\right]\left[1-\exp\left(-na/\pi r^2\right)\right]. \tag{2.20}$$

The first factor in (2.20) is 0 if $r = 0$, while the second is 0 as $r \to \infty$, so there must be a maximizing value for r. The value turns out to be $r^* = \sigma(4z)^{1/4}$, where $z = na/(2\pi\sigma^2)$, as can be verified by showing that $(d/dr)p(r^*) = 0$. Upon substituting r^* into (2.20), one obtains the strategically uniform locally random (SULR) formula

$$P_K = p(r^*) = (1-\exp(-\sqrt{z}))^2. \tag{2.21}$$

Formula (2.21) also holds when $\sigma_U \neq \sigma_V$, provided that $z = na/(2\pi\sigma_U\sigma_V)$ and that the confetti is scattered uniformly over an optimally sized ellipse. Figure 4 shows that the SULR formula (2.21) provides a much smaller estimate of P_K than does (2.19).

The final approximation is the same confetti approximation, except that the coverage ratio is allowed to be any function $d(x, y)$ of two spatial coordinates, subject of course to being nonnegative and to the constraint that the total amount of confetti used must be na. This includes the case where $d(x, y)$ is constant within some region and 0 outside it, so we should expect the current approximation to be larger than (2.21). This is the strategically optimal, locally random (SOLR) case. Formally, the optimization problem is

$$\begin{aligned} &\text{maximize } \iint f(x,y)\left[1-\exp\left(-d(x,y)\right)\right]dx\,dy \\ &\text{subject to } d(x,y) \geq 0 \text{ for all } x, y \\ &\qquad \text{and } \iint d(x,y)\,dx\,dy = na, \end{aligned} \tag{2.22}$$

where $f(x, y)$ is the bivariate normal density function with standard deviations (σ_U, σ_V). The solution can be found in Morse and Kimball (1950, Chapter 5), together with a discussion of how the optimal coverage ratio function $d^*(x, y)$ can be used as a guide in designing effective patterns. The optimal function $d^*(x,y)$ is

$$d^*(x,y) = \frac{1}{2}\left(\sqrt{8z} - \frac{x^2}{\sigma_U^2} - \frac{y^2}{\sigma_V^2}\right)^+, \tag{2.23}$$

where the + indicates that $d^*(x,y)$ is to be 0 rather than negative, and where as usual $z = na/(2\pi\sigma_U\sigma_V)$. Note that the confetti should be most dense at the origin, with the density falling off gradually to 0 on the $(8z)^{1/4}$ -standard deviation ellipse. Outside of this ellipse there should be no confetti at all. The result of substituting $d^*(x,y)$ into the objective function is the SOLR formula

$$P_K = 1 - \left(1 + \sqrt{2z}\right)\exp\left(-\sqrt{2z}\right), \tag{2.24}$$

The SOLR formula is also shown in Figure 4. There is not much difference between (2.21) and (2.24). Once the total lethal area has been conceptually reduced to confetti, it turns out not to be crucial that its distribution be exactly (2.23).

Example 4: Suppose that there are four weapons with cookie-cutter damage functions, with $R = 7.5$, and that the error standard deviations are $\sigma_U = \sigma_V = 7.5$, $\sigma_X = \sigma_Y = 1$. By exhaustive trial and error computations, it can be determined that the exact best pattern is a square of side 11.7, and that the associated kill probability is 0.80. Since $z = 4\pi(7.5)^2/(2\pi(7.5)^2) = 2$, the three approximations are 0.865, 0.594 (SOLR), and 0.573 (SULR). The upper bound is considerably closer to the truth than either of the confetti approximations. The confetti approximations can be made to look better by letting the weapons be diffuse Gaussian with the same lethal area, in which case the approximations do not change but exact computations reveal that the best P_K is only 0.69, achieved by aiming the four weapons in a square of side 10. If the dispersion error is in addition increased from 1 to 5, the approximations still do not change, but the best possible P_K decreases to 0.62.

Since neither σ_X, σ_Y, nor any feature of the damage function other than lethal area enters the computation of z, it is clear that one could find cases where the actual kill probability is even smaller than the confetti approximations. In fact, one has only to consider any problem where the shots are nearly independent, since $z = \infty$ when σ_U or σ_V is 0. In problems where the bias errors dominate the dispersion errors, however, the confetti approximations can usually be thought of as lower bounds on P_K.

There is one other bias-compensation technique that is worthy of mention. Instead of aiming the several shots in a pattern, one can simply aim them all at the target, but inaccurately. A shotgun blast is an example of this; the marksman aims at only one place, but the shotgun pellets form a random pattern around it. Sheet

"SULR_SOLR" of *Chapter2.xls* demonstrates this for the SULR and SOLR approximations by generating the shot locations randomly from the appropriate density, and then testing whether any of the shots actually kills the randomly located target. This is a Monte Carlo simulation (Appendix C). If the deliberate inaccuracy itself takes the form of a bivariate normal error for each shot, the technique goes by the name of "artificial dispersion" and is amenable to analysis because only the sum of artificial and real dispersions is important. Significant work on determining the right amount of artificial dispersion was performed during World War II in Russia (Kolmogorov, 1948). The overall effect should be similar to one of the confetti approximations.

Given all the above considerations, we offer the following procedure for obtaining an approximate P_K in the general case where both bias and dispersion are present:

(a) If dispersion dominates bias, determine the "equivalent" dispersion standard deviations $\sigma'_X = \sqrt{\sigma_X^2 + \sigma_U^2}$ and $\sigma'_Y = \sqrt{\sigma_Y^2 + \sigma_V^2}$ for each shot, and then use (2.13) to obtain an approximate P_K.
(b) If bias dominates dispersion, and if the "packing problem" can probably be solved without much overlap (nearly cookie-cutter weapons, dispersion small compared to lethal radius as well as bias, etc.), use (2.19). This is an upper bound.
(c) If bias dominates dispersion, and if it is clear that the best pattern will involve substantial overlap, use one of the confetti approximations.

The above rules are not exhaustive, since there are certainly cases where neither type of error dominates the other, and in any case the resulting estimate of P_K is only an approximation. An accurate P_K can only be obtained by evaluating (by Monte Carlo simulation, for example – see Appendix C and Exercise 8) sufficiently many patterns to be sure of having discovered the best one.

2.3.3 The Diffuse Gaussian Special Case

The case of the DG damage function with normally distributed errors is exceptional in that P_K can be evaluated analytically for any given pattern of shots, even when the shots have different parameters. As a result, it is possible to evaluate and even optimize a given pattern using a spreadsheet and some VBA code. This is an important enough capability that we include the required mathematics in this section, but the reader may wish to skip it in favor of sheet "DGPattn" of *Chapter2.xls*, which employs the mathematics in the form of VBA code, or to refer to the expanded and slightly generalized treatment in Washburn (2003a).

Formula (2.9) is a product of two factors, one depending on X-parameters and the other on Y-parameters. Let

$$K(b,\mu,\sigma)=\frac{b}{\sqrt{b^2+\sigma^2}}\exp\left(-\frac{1}{2}\frac{\mu^2}{b^2+\sigma^2}\right). \tag{2.25}$$

Then (2.9) states that $P_K = K(b_X,\mu_X,\sigma_X)K(b_Y,\mu_Y,\sigma_Y)$. Now, let n shots be aimed at (x_i, y_i), i=1,...,n. If (U, V) is the bias error common to all shots, then the ith shot is offset from the target by $(\mu_X,\mu_Y)=(x_i+U,y_i+V)$, and the kill probability for the ith shot is therefore $P_i(U,V)=K(b_{X_i},x_i+U,\sigma_{X_i})K(b_{Y_i},y_i+V,\sigma_{Y_i})$. Here we have changed the name of the product because we wish to reserve the symbol P_K for the kill probability of the whole collection of n shots. Given the bias error, the shots are independent and we can employ (2.13), so

$$1-P_K = E(\prod_{i=1}^{n}\{1-P_i(U,V)\}). \tag{2.26}$$

In (2.26), which is an application of the theorem of total probability, the expected value is with respect to the normal distribution of the common bias error (U, V). Note that each shot can potentially have distinct parameters for dispersion and lethality, with b_{Xi} and σ_{Xi} being the lethality and dispersion of the ith shot in the X-direction, etc.

The first step in evaluating (2.26) is to expand the product into a sum of terms, each of which is associated with one of the 2^n subsets of the n shots. For the subset S, the term is

$$t_S \equiv E(\prod_{i\in S}P_i(U,V)). \tag{2.27}$$

The plan is to analytically evaluate each of these terms, and then sum up all 2^n of them, paying proper attention to signs. Our first observation is that, since U and V are independent errors, and since $P_i(U,V)$ can be factored into an X-part and a Y-part, so can t_S. In fact, let

$$t(\mathbf{b},\boldsymbol{\mu},\boldsymbol{\sigma},s,S)\equiv E(\prod_{i\in S}K(b_i,\mu_i+W,\sigma_i)) , \tag{2.28}$$

where $\mathbf{b}$, $\boldsymbol{\sigma}$, and $\boldsymbol{\mu}$ are n-vectors indexed by shot number, and the expectation is taken with respect to a normal distribution of W that has standard deviation s. Then $t_S = t(\mathbf{b}_X,\mathbf{x},\boldsymbol{\sigma}_X,\sigma_U,S)t(\mathbf{b}_Y,\mathbf{x},\boldsymbol{\sigma}_Y,\sigma_V,S)$. Here $\mathbf{b}_X=(b_{X_1},...,b_{X_n})$, and similarly $\boldsymbol{\sigma}_X$, $\mathbf{x}$, $\mathbf{b}_Y$, $\boldsymbol{\sigma}_Y$, and $\mathbf{y}$ are also n-vectors. The central problem is now to evaluate $t(\mathbf{b},\boldsymbol{\mu},\boldsymbol{\sigma},s,S)$, since this will lead to the evaluation of P_K through (2.26).

Let $M_k \equiv \sum_{i \in S} \mu_i^k / (b_i^2 + \sigma_i^2)$; $k = 0, 1, 2$, let $C \equiv \prod_{i \in S} \left(\frac{b_i}{\sqrt{b_i^2 + \sigma_i^2}} \right)$, and recall the definition of $K()$ in (2.25). The product in (2.28) is

$$\prod_{i \in S} K(b_i, \mu_i + W, \sigma_i) = C \exp\left(-\frac{1}{2}(W^2 M_0 + 2WM_1 + M_2) \right) \tag{2.29}$$

The expectation of (2.29) with respect to W can be accomplished analytically because (2.29) involves a quadratic expression in W lying within an exponential. The result is

$$t(\mathbf{b}, \boldsymbol{\mu}, \boldsymbol{\sigma}, s, S) = \frac{C}{\sqrt{1 + s^2 M_0}} \exp\left(-\frac{1}{2}(M_2 - \frac{s^2 M_1^2}{1 + s^2 M_0}) \right). \tag{2.30}$$

Calculating P_K is now just a matter of going through the steps outlined earlier. One can even include a shot reliability by modifying the C-factor. An implementation of the method outlined above is the VBA function *PK*() included with *Chapter2.xls*. The method is essentially that of Bressel (1971), who goes on to take the additional step of averaging over a rectangular target. The *PK*() function uses double-precision arithmetic because the series being summed alternates in sign. See Grubbs (1968) for alternatives with better numerical stability.

Since P_K can be evaluated analytically as a function of the aim point vectors $\mathbf{x}$ and $\mathbf{y}$, we can now consider the problem of finding the pattern that will maximize P_K. This is a nonlinear optimization problem. On sheet "DGPattn" of *Chapter2.xls*, Excel's Solver is employed to accomplish this.

2.3.4 Area Targets and Multiple Error Sources

Section 2.3.2 is often applicable even when there are multiple sources of error. Suppose, for example, that

(a) the location of a target relative to some known datum is E_1.

(b) all shots are to be fired from a platform whose location relative to the same datum is subject to an error E_2.

(c) each shot has an individual firing error E_3 due to trembling on the part of the marksman.

(d) an additional firing error is introduced due to an unknown wind velocity E_4.

(e) E_1, E_2, E_3, and E_4 are all independent, normal random variables with 0 mean and variances σ_i^2, $i = 1, 2, 3, 4$.

It is necessary to classify each of the four errors as either "bias" or "dispersion." E_1 and E_2 are clearly bias errors, since we assume that the positions of the target and the firing platform are the same for each shot. E_3 is clearly a dispersion error, since each shot has an independent dispersion that is different from all the rest. E_4 might be a bias error if the unpredictable part of wind velocity were constant in space over the length of time required to fire the shots (the predictable part is irrelevant, since the marksman could allow for it in aiming) or it might be a dispersion error if the wind were very gusty. Assume the latter. Then, making the natural assumption that the four error types are independent of each other, and noting that it is only the total bias error and the total dispersion error that affect the fate of the target, the equivalent bias and dispersion error variances are $\sigma_1^2 + \sigma_2^2$ and $\sigma_3^2 + \sigma_4^2$, respectively, and Section 2.3.2 can be applied to the equivalent errors. The principle being used is the theorem that the variance of a sum of independent random variables is the sum of the variances.

All of the above applies to targets whose only property is a location; i.e., to point targets. However, it is not difficult to handle area targets within this scheme, provided we are interested only in the average fraction of the target killed. In (2.18), if $f(u,v)$ is interpreted to be the amount of target value per unit area at point (u,v), then P_K has the meaning "total amount of value killed, on the average." Furthermore, if we normalize $f(u,v)$ so that the total target value is unity, then P_K means "average fraction of the target killed," or, equivalently, the probability of killing a test element of the target, where the location of a test element is governed by the density function $f(u,v)$. In other words, the test element location (U,V) can be regarded as just another bias error, to be combined with any other bias errors as necessary. Any area target can be handled by converting the value density of the area target to an equivalent density function of a bias error, and then proceeding as if the target were a point target. This is especially easy to do, of course, if $f(u,v)$ happens to be bivariate normal.

Example 5: Suppose that (U, V) is circular normal with standard deviation 80 ft; that is, assume that the target's value is spread out in a bell-shaped manner. Also assume that E_1, E_2, E_3, and E_4 are all circular normal with standard deviations 10, 20, 30, and 40 ft, respectively. Assuming that the wind error is dispersion, the equivalent dispersion is $\sigma_X = \sigma_Y = \sqrt{30^2 + 40^2} = 50$ ft, and the equivalent bias is $\sigma_U = \sigma_V = \sqrt{10^2 + 20^2 + 80^2} = 83$ ft. One could now proceed as in Section 2.2, probably by ignoring the dispersion error and using the SOLR formula to estimate P_K, which is now interpreted as the maximum possible expected fraction of the target killed by an optimal pattern. If the pattern consists of 20 shots with lethal radius 10 ft each, we find that $z = 20(10)^2/(2(83)^2) = 1/69$. The resulting P_K is very small, even according to the optimistic (2.19). Twenty shots of this size are simply not capable of doing much damage to a target as spread out as the one postulated.

The fact that area targets introduce an effective bias error is important in determining whether CMP or EMT is a better measure of effectiveness for an arsenal of weapons (see Section 2.3.1). Since (2.14) was derived under the assumption that the only firing error was dispersion, we can say that CMP is the proper measure if the targets are point targets and if the bias errors are negligible. If the effective bias (including the effects of target size) dominates the effective dispersion, however, then EMT is more appropriate. Thus (to conclude the comparison that was begun in Section 2.3.1), the United States nuclear arsenal in 1978 was more effective against well located, hard targets such as ICBM silos, but the Soviet Union arsenal was more effective against cities, which are well-located area targets, or against submarines, which are poorly located point targets. Dispersion is almost irrelevant for either of these latter target types, even though it is crucial for the former.

2.4 Multiple Shots, Multiple Targets, One Salvo

In this section we consider situations where there are multiple targets, all of which must be attacked "simultaneously" in the sense that no feedback about results is available between shots (the next chapter considers feedback). If several shots are available, the question of how they ought to be distributed over the targets arises.

2.4.1 Identical Shots, Identical Targets, Optimal Shooting

If all shots in a salvo are identical and all targets are identical, then any reasonable measure of effectiveness will be maximized when the shots are spread as evenly as possible over the targets. If the number of shots is an integer multiple of the number of targets, then the number of surviving targets is a binomial random variable, since all targets have the same survival probability. The situation is somewhat more complicated when the number of shots is not an integer multiple of the number of targets, since some targets are shot at one more time than others. Suppose there are b shots and n targets, with each shot having a miss probability q against its target. Let k be the integer part of b/n, so that some targets are shot at k times while others (r of them) are shot at $k+1$ times; r is the number of shots left over after every target is shot at k times, so $r = b - kn$. The total number of surviving targets is the sum of two binomial random variables, the number of survivors from the r targets shot at $k+1$ times and the number of survivors from the n–r targets shot at k times. The distribution of the total number of survivors X is therefore the convolution of two binomial distributions. The probability mass function of X is given by

$$Surv(x;b,n,q) \equiv \sum_{j=\max(0,x-n+r)}^{\min(r,x)} Bin(j;r,q^{k+1})Bin(x-j;n-r,q^{k}), \tag{2.31}$$

where $Bin(y;t,p)$ is the binomial probability of y successes out of t trials when the success probability is p. The average number of survivors is

$$E(X) = rq^{k+1} + (n-r)q^{k}. \tag{2.32}$$

The function *Surv*() is included in module 1 of *Chapter2.xls.*

Example 6: Suppose that $n = 3$ targets are attacked by $b = 4$ weapons, each of which has an individual miss probability of $q = 0.6$. Then $k = 1$ and $r = 1$, so one target is attacked twice while the rest are attacked only once. The probability of three survivors is $q^4 = 0.1296$, since all four shots must miss if all targets are to survive (the sum in (2.30) contains only the term for $j = 1$). The rest of the probability distribution of X can best be obtained using the *Surv*() function defined in (2.30). From (2.31), the average number of survivors is 1.56. See sheet "Surv" of *Chapter2.xls*.

2.4.2 Identical Shots, Diverse Targets, Optimal Shooting

Here we consider the case where the shots are all identical, but not the targets. This kind of situation is roughly the case in planning a nuclear first strike, so considerable work was done on the problem during the Cold War.

When targets are diverse, there are many possible measures of effectiveness. We will assume that each target has a given value, and that the marksman's goal is to kill as much value as possible, on the average. Determining target values is in practice problematic, but nonetheless necessary if an optimal attack is to be planned – only scalar measures can be optimized, and assigning values is the most direct way to put all targets on the same scale. Let v_j be the value of target j, and let q_j be the probability that one shot at target j will fail to kill it (the kill probability is $1-q_j$).

Since all shots are assumed to be independent, the probability that x_j shots will fail to kill target j is $q_j^{x_j}$. If there are n targets and b shots in total, this leads to the problem of minimizing the total average surviving value, subject to not using more shots than are available, which we name problem P1:

$$\text{Minimize} \sum_{j=1}^{n} v_j q_j^{x_j}, \tag{2.33}$$

subject to $\sum_{j=1}^{n} x_j \le b$, and all variables x_j must be nonnegative integers.

Problem P1 has a nonlinear objective function, but nonetheless turns out to be an easy optimization. It can be solved by a "greedy" technique where shots are allocated sequentially, with each shot being allocated to the target for which the incremental target value killed is maximal. Imagine that the problem has already been solved, with **x** being the vector of optimal allocations, and that one more shot becomes available. If the shot is allocated to target j, the decrease in surviving value will be $v_j q_j^{x_j} - v_j q_j^{x_j+1} = v_j q_j^{x_j}(1-q_j)$. The greedy technique starts out with $\mathbf{x} = 0$, and then follows the principle "always assign the next shot to the target for which the decrease in surviving value is largest." The procedure is efficient, and can easily find optimal allocations when targets and shots number in the thousands.

Example 7: Suppose there are three targets, all with value 1, and $\mathbf{q} = (0.1, 0.5, 0.9)$. Using the greedy method, the first 4 shots should be at targets 1, 2, 2, and 2, with the total surviving value after 4 shots being $0.1 + 0.125 + 0.9 = 1.125$. Note that most shots go to the target that is neither too easy nor too hard to kill. Target 1 is so easy to kill that a single shot will suffice, and target 3 is so hard to kill that shots against it are nearly wasted. The fifth shot, however, should be on either target 1 or target 3, since either shot will reduce the surviving value by 0.09, which exceeds the 0.0625 reduction available from target 2.

2.4.3 Diverse Shots, Diverse Targets, Optimal Shooting

In the 1991 Gulf War, the allies were confronted with the problem of attacking Iraq's air defense system in a single, coordinated attack. An air defense system consists of multiple targets, and weapons of various kinds (aircraft, helicopters, cruise missiles) were available to attack it. The planners of that attack had to decide which weapons should attack which targets, and to do so without the prospect of getting much information feedback, since it was imperative that the attack happen quickly. This section deals with such situations, but one important aspect of that attack will be missing in this chapter. The actual attack anticipated losses to the attackers, with the losses depending very much on who attacked what. Such two-sided models will not be considered until Chapter 4.

Since the shots are diverse, the miss probability now depends on both the target and the weapon. Let q_{ij} be the probability that a weapon of type i will miss target j if assigned to attack it. All attacks are assumed to be independent, so powering up still applies. If x_{ij} weapons of type i are assigned to attack target j, the miss probability is therefore $\prod_i q_{ij}^{x_{ij}}$. The object is now to solve problem P2 defined below:

$$\text{Minimize} \sum_j v_j \prod_j q_{ij}^{x_{ij}}, \tag{2.34}$$

subject to $\sum_{j=1}^{n} x_{ij} \le b_i$ for all weapons i, and all variables nonnegative integers.

The greedy method of always making the next assignment to reduce the total surviving value as much as possible does not work on problem P2, as shown in the following example:

Example 8: Suppose there are 2 targets, each with value 1, and 2 weapons. Weapon A has miss probabilities of 0 and 0.1 on the 2 targets, while weapon B has miss probabilities of 0.1 and 0.5. The greedy method would first assign A to target 1, since A will certainly kill it, and would then assign B to target 2. The total surviving value would be 0 + 0.5 = 0.5. The optimal allocation is just the reverse – if B is assigned to 1 and A is assigned to 2, the surviving value is only 0.1 + 0.1 = 0.2. Weapon A is more effective than weapon B against all targets, and the greedy mistake is to lose sight of all other facts in assigning weapon A. A realistic version of this example might have cruise missiles playing the role of weapon A. Effective but expensive weapons such as cruise missiles should be used on high-value targets that are hard for other weapons to kill, so one needs to have a global perspective in assigning them.

Problem P2 is sometimes referred to as the weapon target assignment problem. It is known to be of a fundamentally difficult type, so no simple method like the greedy method will be able to solve it. Optimal methods can be expected to be lengthy on large problems, but a variety of efficient approximations and bounds have been investigated (Ahuja et al., 2003; Washburn, 1995a).

2.4.4 Identical Shots, Identical Targets, Random Shooting

Combat sometimes happens fast enough or communications are sufficiently difficult that deliberate attempts to jointly optimize fires are impossible. Here we investigate the consequences if several identical shots are randomly aimed at several identical targets, rather than optimally aimed as in Section 2.4.1. The assumption of randomness reflects the idea that coordination is impossible, so that a marksman is as likely to shoot at one target as another, independent of the other marksmen.

Thomas (1956), in the context of a situation where several bombers are suddenly ambushed by a group of interceptors, derives the probability distribution of the number of surviving bombers. Using the notation of Section 2.4.1, let $RSurv(x;b,n,q)$ be the probability that x targets survive when b shots with individual miss probability q are randomly distributed over n targets. Thomas shows that the probability mass function of X, the number of survivors, is given by

$$RSurv(x;b,n,q) \equiv \binom{n}{x} \sum_{j=0}^{n-x} (-1)^j \binom{n-x}{j} \left(1 - \frac{(1-q)(x+j)}{n}\right)^b ; x = 0,\ldots,n \qquad (2.35)$$

The function *RSurv*() is also provided in *Chapter2.xls*. See Exercise 13.

The expected value of X is comparatively easy to derive. Consider any particular target. That target will be killed by any particular shot with probability $(1 - q)/n$, since the target is chosen by the shot with probability $1/n$. Since there are b shots, each of which has an independent chance of killing the target, we can power up to find the target's survival probability, which is $(1-(1-q)/n)^b$. Now, let I_x indicate whether target x is killed ($I_x = 1$) or not ($I_x = 0$), so that $X = I_1 + \cdots + I_n$. We know that $E(I_x) = (1-(1-q)/n)^b$ for every x. Since expected values and sums commute, even when the summed random variables are dependent, as they are here, the expected number of survivors is

$$E(X) = n(1 - \frac{1-q}{n})^b. \qquad (2.36)$$

A comparison between random and uniform allocation of shots is of interest for the case where all shots and targets are otherwise identical. We expect that (2.32) will show a larger average number of survivors than (2.36), since surely there is no advantage to the marksman for aiming randomly. This turns out to be true. When $q = 0.6$ and $n = 100$ (or any large number), the average survival probability $E(X)/n$ is shown in Figure 5 for varying numbers of shots per target (k). The curves differ, but by surprisingly little, especially when k is either small or large. See sheet "OptRand" of *Chapter2.xls* to experiment with changes to n or q.

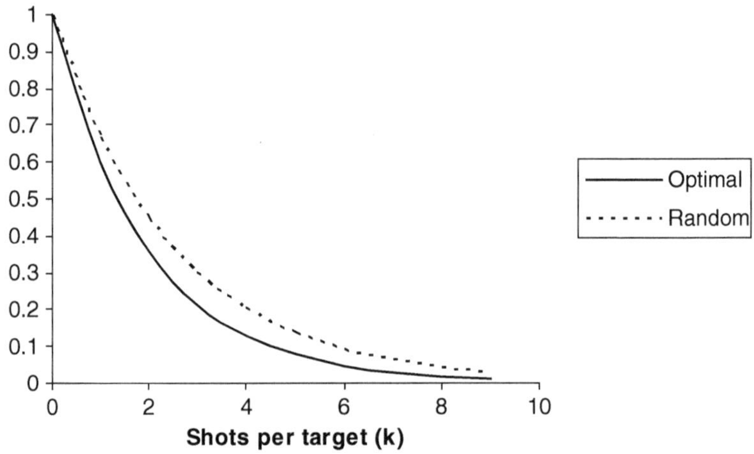

Figure 5: E(X)/n for optimal and random shooting, $q = 0.6$.

2.5 Further Reading

A great deal of work on aiming and kill probability computation was inspired by World War II. Morse and Kimball (1950, Chapter 5) record some methods used in the United States during that war. Wartime Soviet work on artificial dispersion and other topics can be found in Kolmogorov (1948). Eckler and Burr (1972) is a good summary of subsequent work up to 1972, or see Przemieniecki (2000), which includes some basic computer programs for computing kill probability.

Although this chapter emphasizes firing problems in two dimensions, there are also one-dimensional problems of interest. Morse and Kimball (1950, Chapter 5) consider the problem of bombing a railroad track, and other linear structures such as roads or power lines may also prompt analysis. One advantage of one-dimensional problems is that the two-dimensional difficulty of packing the plane with circles disappears. With cookie-cutter weapons, in the absence of significant dispersion errors, the best pattern in one dimension is typically a “stick” that covers an interval with no gaps. Morse and Kimball observe this, and David and Kress (2005) generalize to situations where the bombs are heterogeneous and asymmetric in their effects.

The first few sections of this chapter treat firing errors as being one of two types. The first type is where all firing errors are mutually independent, and the second type adds the possibility that there may be a common component. Neither of these assumptions fit well for rapid-fire weapons, where the sequence of errors is better viewed as a stochastic process. Work on applications of such processes was inspired by aerial combat in World War II in Great Britain, and subsequently reported in Cunningham and Hynd (1946). See also Fraser (1951, 1953).

There is much in common analytically between the problem of firing at a set of diverse targets and the problem of trying to detect a stationary target that is lost among a set of cells. Just as shots are allocated to targets with the object of killing them, units of search effort are allocated to cells with the object of detecting the lost target. Problem P1 in Section 2.4.1 serves both points of view and has benefited analytically from analysts in both camps. Perusal of the part of search theory literature devoted to stationary targets may prove fruitful to someone interested in firing theory. See Stone (1975), for example.

All military services include killing targets among their duties, and many weapons are shared among the services. It therefore makes sense to have a joint organization responsible for determining how to calculate kill probabilities. In the United States, this is the Joint Technical Coordinating Group for Munitions Effectiveness, or JCTG/ME. The principal products of the JTCG/ME are Joint Munitions Effectiveness Manuals (JMEMs), some of which are in the form of standardized computer programs. JMEMs employ some of the methods covered earlier in this chapter, but also other methods that are more computer intensive and include more detail. These manuals (there are hundreds of them) also include classified data for specific weapons and targets. At the unclassified level, a good way to find out more about JCTG/ME or JMEM is to go to the Federation of American Scientists (FAO, 2008) and search on either of those acronyms.

Exercises

(1) Suppose $D(r) = 1 - r$ if $r < 1$; 0 if $r > 1$. What is the lethal area?
Ans. $a = \pi/3$

(2) Plot $D(r)$ for the target illustrated below, assuming that the weapon must hit the shaded area and that the impact point is (r, θ) with θ uniformly random in $[0, 2\pi]$. Show that the lethal area is equal to the area of the target.
Ans. $D(r)$ is a step function, $a = 2.5\pi$

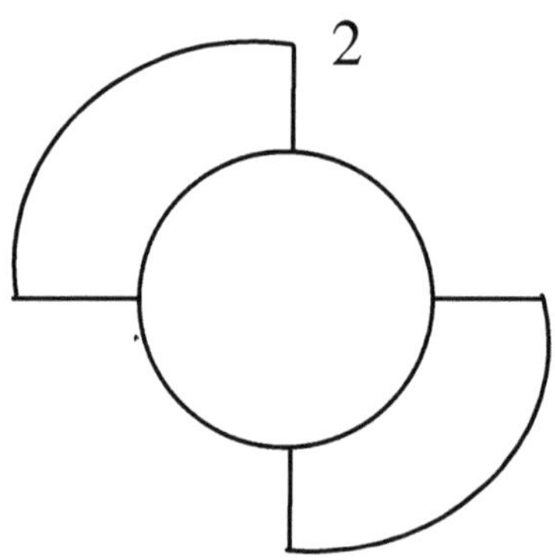

(3) Show that (2.4) produces $\pi E(R^2)$ for lethal area, where $E(R^2)$ is computed using the density function given by (2.6). Hint: use integration by parts.

(4) Derive (2.7).

(5) When aiming errors are basically angular, the miss distances should increase with range. Suppose several independent shots are taken at a target, with $\sigma_i = 0.1r_i$, where r_i is the ith range, and that the cookie-cutter lethal radius is 1. If the successive ranges are 10, 11, 12, …, increasing by unity with each shot, compute P_K for the first shot, the first five shots as a group, and the first 10 shots as a group.
Ans. ($P_K\,(1) = 0.39$, $P_K\,(5) = 0.84$, $P_K\,(10) = 0.93$).

(6) A weapon with a radially symmetric DG damage function is aimed at a terrorist camp that is located 100 m from a baby-milk factory. The weapon has different lethalities for camp and the factory. The parameter b (the common value of b_X and b_Y) is 50 m for the camp and 80 m for the factory. The circular normal firing error has standard deviation 50 m. What are the probabilities of killing the camp and the factory?
Ans. (0.50 for the camp, 0.41 for the factory)

(7) An aircraft attempts to kill a tank as follows: It first drops a canister of "stickers" in the hope that one will hit the tank and activate. If a sticker acti-

vates, it can guide a projectile to the tank. The canister opens and scatters 1000 stickers, with the amount of scatter being under the control of the designer. The exposed area of the tank is 100 m^2. The aircraft makes a two-dimenslonal error with standard deviations (100 m, 300 m) in dropping the canister. What is the probability that a sticker hits the tank, assuming a well-designed canister? If the tank is longer than it is wide, does the direction of the aircraft's approach matter?
Ans. (0.275, no)

(8) Suppose you are given 16 detection devices, each of which is guaranteed to detect a target if and only if the relative distance is either less than 4 miles or between 30 and 33 miles (the "convergence zone" phenomenon in the ocean might be one explanation for such an assumption). The devices can be placed in any pattern whatever, and the object is to detect a target whose location relative to some known point is circular normal with standard deviation 30 miles in each direction. There are no dispersion errors.
   (a) Estimate p_K.
   (b) Make up a pattern and test it by writing a 5000-replication computer simulation.

Ans. The lethal area is $\pi(4^2 + 33^2 - 30^2) = 205\pi$ square miles, so $z = (16)(205)/1800 = 1.82$. Given that the shape of the lethal area makes considerable overlap inevitable even in the absence of dispersion, a confetti approximation is natural. The SOLR formula produces $P_K \approx 0.57$. This example has been the result of considerable experimentation, with the best pattern as of this writing having a detection probability of 0.64. The exact answer is unknown.

(9) Suppose 10 cookie-cutter shots are available, with the lethal radius being 30 ft for each. Estimate P_K for the area target and errors considered in Section 2.3, assuming that
   (a) the wind error is dispersion
   (b) the wind error is bias

Ans. Using the SOLR formula in both cases, the expected fraction of the target killed with an optimal pattern would be approximately 0.316 in case a), or 0.275 in case (b).

(10) Use sheet "DGPattn" of *Chapter2.xls* to verify the two DG claims made in Example 4. Hint: To make the weapons have the right lethal area, you must set the DG lethality parameter b so that $2\pi b^2 = \pi(7.5)^2$.

(11) Consider problem P1 of Section 2.4.2, with $n = 3$, $\mathbf{v} = (1,2,3)$, and $\mathbf{q} = (0.3, 0.5, 0.8)$. Allocate the first four shots using the greedy algorithm.

Ans. The shots should be in the order 2, 1, 3, 2, and the average value surviving should be 3.2.

(12) Write a computer program to solve problem P1 of Section 2.4.2, using the greedy algorithm. Use it to solve Example 7.

(13) Insert a new worksheet into *Chapter2.xls* and use it to analyze a situation where there are $n = 3$ targets and $b = 4$ shots, each of which has a miss probability of $q = 0.6$.

(a) Use the *Surv*($x;b,n,q$) function to compute the probability distribution of the number of survivors when b shots with miss probability q are distributed evenly over n targets. Use that distribution to compute the mean number of survivors, and verify that (2.31) gives the same result.

(b) Use the *RSurv*($x;b,n,q$) function to compute the probability distribution of the number of survivors when b shots with miss probability q are distributed randomly over n targets. Use that distribution to compute the mean number of survivors and verify that (2.35) gives the same result.

Ans. A partial answer is that the mean number of survivors should be 1.56 and 1.69 in the even and random cases, respectively. See sheet "Surv"of *Chapter2.xls* for a solution.

(14) If formulas 2.7 and 2.8 are both true, show that CEP must be related to σ as claimed in Section 2.2.2.

(15) Consider problem P2 of Section 2.4.3, with $\mathbf{v}$ = (3,1) and $\mathbf{q} = \begin{bmatrix} 0.5 & 0.0 \\ 0.6 & 0.5 \\ 0.7 & 0.5 \end{bmatrix}$.

There is a single weapon of each of the three types, and you are to assign weapons to maximize the total value killed, or equivalently to minimize the total surviving value. The best that can be done with one weapon is to assign weapon 1 to target 1, since this will kill 3(0.5) units of target value. Complete the application of the greedy algorithm to this problem, find the optimal assignment, and compare the two.

Ans. The greedy algorithm results in weapons 1 and 3 assigned to target 1, with a total surviving value of 3(0.5)(0.7) + 1(0.5) = 1.55. The optimal assignment has weapons 2 and 3 assigned to target 1, with a total surviving value of 3(0.6)(0.7) + 1(0) = 1.26.

(16) Page "Shotgun" of *Chapter2.xls* is a Monte Carlo simulation of a shotgun with 50 pellets trying to hit a target in the face of bias error. Relative to the bias, each pellet hits independently with a uniformly distributed angle and a dispersion distance X whose cumulative distribution function is $F(x) = 1 - \exp(-(x/b)^3)$. This is a Weibull distribution, which is often realistic and always easy to simulate because it is possible to solve the equation for x. Now suppose that the only control you have over the pattern is through the dispersion parameter b. This control might be exercised through a choke on a shotgun or through the dispersement altitude if the pellets are actually bomblets. Experiment with the sheet to find the best value of b when the bias

standard deviation is 50 and the lethal radius is 10 and compare the resulting kill probability with the SOLR formula. Use *SimSheet.xls* to replicate the experiment a large number of times.
Ans. With b set to 50, 60, 80, and 100, the kill probability after 3000 replications is approximately 0.376, 0.416, 0.393, and 0.328, respectively. The best setting for b is about 60, and the SOLR formula comes close to predicting the optimized kill probability.

(17) The DG damage function (2.9) can be reduced to one dimension (the x-dimension, say) by setting $\mu_Y = \sigma_Y = 0$, so Section 2.3.3 and its accompanying sheet "DGPattn" of *Chapter2.xls* also apply to finding multiple-shot patterns in one dimension. Use that sheet to find the optimal placement of four identical bombs with unit reliability, $b_X = 30$ and $\sigma_X = 30$. Assume that the common error has standard deviations of $(\sigma_U, \sigma_V) = (100,0)$; that is, there is no bias error in the vertical dimension. In using that sheet, the value of b_Y is irrelevant as long as it is not 0.
Ans. The optimized pattern is (–100, –30, 30, 100), with a kill probability of 0.68. You will not discover that fact if you start at (0, 0, 0, 0), since all derivatives are 0 there and Solver will therefore not move the aim points, so use some other starting point. If you increase the dispersion to $\sigma_X = 100$ and re-optimize, all four bombs should be aimed at the origin. This is typical when the dispersion error is large.

Chapter 3
Shooting with Feedback

I didn't know the gun was loaded,
And I'm so, so sorry, my friend.
I didn't know the gun was loaded,
And I'll never, never do it again.

(lyrics by Hank Fort and Herb Leighton, 1949)

3.1 Introduction

This chapter deals with shooting situations where there is information feedback between shots and salvos of shots. The effect of this information is to decrease the number of shots required, to increase the kill probability, or even (as in the introductory quote) to discover something about the weapon doing the shooting. In spite of the implied extended time period, we assume as in Chapter 2 that shooting is one sided, with no return fire. Although problems with multiple shots are also considered in Chapter 2, the addition of information feedback is an essential change to the nature of the problem, as will be seen.

The information feedback can take various forms. One is a report about the effect of previous shots on the target. For example, the same radar that keeps track of the target's location might also be able to detect when the target breaks up into smaller pieces, thus indicating target destruction. The target's behavior might also change after it is hit. Except for Section 3.3, all problems considered in this chapter are of this sort, with the feedback generally being whether the target is alive or dead.

Another kind of feedback provides information about miss distances, rather than the live/dead status of the target. Artillery registration ("zeroing in") is one example of this, and another is the Close In Weapon System (CIWS), a high rate of fire gun designed to protect ships from air attack. The CIWS radar measures the location of the stream of bullets, as well as the target's location, and so can provide information about the relative locations of shots and target. This information can be the basis of a procedure for adjusting the aim point, thus gradually eliminating bias errors. We consider such problems in Section 3.3.

Some important tactical questions arising during the shooting sequence are

- When do we stop shooting?
- Where do we aim?
- At which target do we aim?

The answers will depend on tactical constraints, as well as feedback information available. There are situations (CIWS, for example) where both target location and

A. Washburn, M. Kress, *Combat Modeling*, International Series in Operations Research & Management Science 134, DOI 10.1007/978-1-4419-0790-5_3,

status are fed back, and target status can be more complicated than a simple live/dead report. Feedback can be subject to error. Each unique situation deserves its own special analysis, so there is no hope of being comprehensive. The sections below describe a few basic situations, but there are many others.

3.2 Feedback on the Status of a Single Target

The shoot-look-shoot (SLS) procedure involves feedback about whether the target has been killed, which in this section is assumed to be perfectly accurate. The general idea is that one shoots, looks to see whether the target has been killed, and shoots again if necessary, possibly repeating this procedure several times. The advantage of such information is that it helps to prevent the assignment of additional shots to a target that is already dead.

The reduction in shots required when using SLS can be significant. For example, suppose that an important communications station is to be attacked by an expensive missile prior to an attack by land forces. The kill probability is 0.9 per shot, but the station is so important that $P_K = 0.99$ is required. By powering up, we see that this can be achieved without feedback by simply assigning two independent missiles to the station. If a look is possible between shots, however, then 90% of the time a single shot will suffice, since the first shot will kill the target with that probability. The second shot is required only 10% of the time, so the average number of shots required with SLS is 1 + (0.1)(1) = 1.1, instead of 2. The SLS procedure requires just over half of the resources that would be required without feedback and has exactly the same kill probability.

In the extreme case where the number of looks is unbounded, the marksman can even adopt the strategy "fire until the target has been killed." In that case the problem is not to compute P_K (which is 1.0), but rather to investigate the random variable $N \equiv$ "number of shots required to kill the target." If the shots are all independent with kill probability p, then N is a geometric random variable (Appendix A) with expected value $E(N) = 1/p$. More generally, if q_i is the miss probability of the ith in a sequence of independent shots, then

$$E(N) = \sum_{n=0}^{\infty} P(N > n) = 1 + q_1 + q_1 q_2 + q_1 q_2 q_3 + \ldots. \tag{3.1}$$

This general formula reduces to a geometric series with sum $1/p$ in the special case where $q_i = 1 - p$ for all i. The first equality in (3.1), although not the usual definition of expected value, can be shown to be valid for any nonnegative integer random variable such as N.

A similar problem is one where the marksman has n shots available, but they all have different costs c_i, as well as different kill probabilities $p_i = 1 - q_i$, for $i = 1, ..., n$. It is not unusual to have considerable diversity in both cost and effectiveness between shot types, especially when the marksman has both guided and unguided weapons available. The target is assumed to be so important that the marksman will shoot at the target until either it is killed or the marksman runs out of shots. The overall miss probability is the product $q_1q_2,...,q_n$, regardless of the order in which the shots are taken. However, the average cost of shooting depends on that order.

Example 1: Suppose there are two shots, a missile and a bomb, with $\mathbf{c} = (2,1)$ and $\mathbf{p} = (1, 0.6)$. If the missile is used first, then the cost will be 2, since its cost is 2 and it will certainly kill the target. If the bomb is used first, then the cost will be 1 with probability 0.6, or 1 + 2 if the bomb misses, for a total average cost of $0.6(1)+0.4(3)=1.8$. Since 1.8 is smaller than 2, as long as the time required to kill the target is not an issue, it is better to use the bomb first.

It turns out that the optimal tactic is always to rank the shots in decreasing "bang-per-buck" order p_i/c_i, and then make the shots in that order. In Example 1, the ratio scores for missile and bomb are 0.5 and 0.6, respectively, so the bomb ranks higher and should be chosen first. See Glazebrook and Washburn (2004, Theorem 4) for a general proof of this statement. If large inventories of each shot type were present, then the optimal tactic would be to choose the highest ranking shot type and use it repeatedly until either it is exhausted or the target is killed. The average cost of killing the target with repeated use of shot type i is c_i/p_i, the product of the cost per shot and the expected number of shots. That ratio is also the reciprocal of the bang-per-buck index, so this conclusion should not be surprising.

Example 2: Suppose there are three shot types, with $\mathbf{c} = (1,2,3)$, $\mathbf{p} = (0.2,0.5,0.8)$, and with two shots of the second type and one of the other two. Since type 3 ranks higher than 2, and 2 ranks higher than 1, the shots should be taken in the order 3, 2, 2, 1. The overall kill probability is $1-(0.2)(0.5)^2(0.8)=0.96$. The average cost of shooting is $3\times0.8+5\times0.2\times0.5+7\times0.2\times0.5^2+8\times0.2\times0.5^2=3.65$. Taking the shots in any other order would result in the same kill probability and a higher average cost of shooting. Without applying SLS, the same overall kill probability would cost 8, since all four weapons would be required.

3.3 Feedback on Miss Distances

In this section, we assume that a forward observer provides feedback to the marksman on the location of each shot's impact point, relative to the target, as in Figure 1. If the marksman is initially uncertain about the target location, the alignment of his own sights or any other source of bias error,this information can be used to "zero-in" on the target by adjusting the aim point. The essence of the problem is to adjust without over-adjusting – the marksman must also consider that there are reasons other than bias for missing the target, notably the independent dispersion errors that accompany the shots, for which adjustment of the aim point would be unwise.

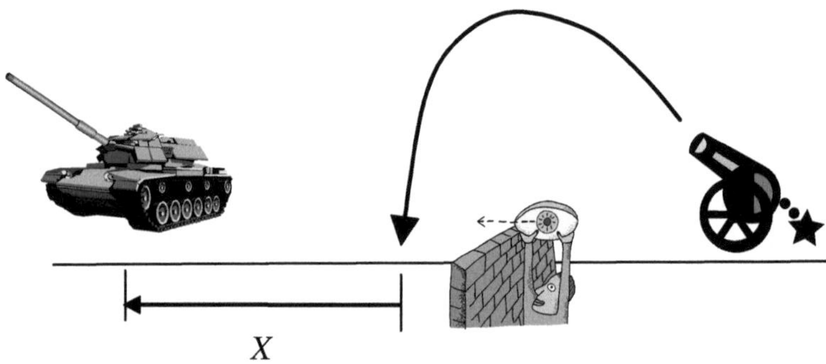

Figure 1: The miss distance X is observed and fed back to the cannon, which can adjust its aim point for the next shot.

It is easiest to begin with an analysis in one dimension. In the shoot-adjust-shoot (SAS) procedure, it is assumed that an observer provides the actual impact point of the ith shot, relative to the target. The observer reports are useful because they help the marksman to estimate whatever bias error B is present, and thereby to adjust the next aim point A_{i+1} to take account of it. Assuming that the dispersion error is E_i for the ith shot, the observer's ith report is

$$X_i = B + E_i + A_i; i > 0. \tag{3.2}$$

The aim point A_i is known and X_i can be observed, but B and E_i are unknown, except for their sum. The aim point A_{i+1} should be the negation of the marksman's minimum variance estimate of the unknown bias B, based on the observed miss distances $X_1,\ldots, X_i$. Since $B + E_i$ is an unbiased observation of B, and since $B + E_i = X_i - A_i$, the next aim point is obtained by averaging:

$$-A_{i+1} = \frac{1}{i}\sum_{j=1}^{i}(X_j - A_j) = B + \frac{1}{i}\sum_{j=1}^{i} E_j; \quad i > 1. \tag{3.3}$$

Therefore, with all aim points but the first being given by (3.3),

$$X_{i+1} = E_{i+1} - \frac{1}{i}\sum_{j=1}^{i} E_j; \quad i > 1. \tag{3.4}$$

Assuming that the dispersion errors are normal, independent, identically distributed random variables with mean 0 and variance σ^2, (3.4) implies that $E(X_{i+1}) = 0$ and, since the variance of a sum of independent random variables is the sum of the variances (see Appendix A),

$$Var(X_{i+1}) = \sigma^2 + (-1)^2 \left(\frac{1}{i}\right)^2 \left(\sigma^2 + \ldots + \sigma^2\right) = \sigma^2 + \frac{\sigma^2}{i} = \sigma^2 \left(1 + \frac{1}{i}\right). \tag{3.5}$$

Equation (3.5) gives the variance of the miss distance for every shot except the first, which we regard as a "calibration shot" that is incapable of killing the target (making $Var(B)$ infinite would have the same effect). Since the miss distances after the first can be shown to be independent of each other, the probability of kill with a fixed number of shots can be obtained with the same independence argument that leads to (2.13).

Equations (3.3) and (3.4) together imply that

$$A_{i+1} - A_i = -X_i / i; i > 0. \tag{3.6}$$

Equation (3.6) states that the aim point for the next shot should be corrected by a decreasingly small fraction of the previous miss distance. Note that early miss distances are taken more seriously than late ones; the marksman is not inclined to make large aim point adjustments late in the firing procedure because the bias has almost been eliminated. Barr (1974) makes the point that the normality assumptions made here are not really necessary for the optimality of this aim point correction rule.

It would be simpler to adopt an aim point correction rule, something like a procedure that we will call 3CAL: aim the first three shots at the target, thinking of them as calibration shots, and then let the aim point for all following shots be $-(X_1 + X_2 + X_3)/3$, the negation of the average of the reported miss distances. Does the simplicity of 3CAL compensate for its smaller effectiveness? Clearly the answer is no if there are only three shots, since the three calibration shots will never hit the target if the bias is large. The answer can therefore be expected to depend on the number of shots. Since the two competing procedures are well defined, answering this question would be a good application for Monte Carlo simulation. Sheet "3CAL" of *Chapter3&4.xls* is such a simulation (see Exercise 8).

Suppose now that the SAS procedure is carried out independently in each of two dimensions, using n cookie-cutter shots with lethal radius R, including the calibration round. The two-dimensional miss distances will then be circular normal with variance given by (3.5), and therefore, using (2.7) and (2.13) in the same manner as in Chapter 2,

$$P_K = 1 - \exp\left(-\frac{R^2}{2\sigma^2}\left(\frac{1}{2} + \frac{2}{3} + \cdots + \frac{n-1}{n}\right)\right). \tag{3.7}$$

Note that the effectiveness (CMP – see Section 2.3.1) of the ith shot, compared to its effectiveness in a problem with no bias error, is $(i-1)/i$. The SAS procedure is evidently not completely effective in getting rid of the effects of the bias error, except in the limit when there are many shots. There is nonetheless a reasonable sense in which it is the optimal aim adjustment procedure (Nadler and Eilbott, 1971; Barr, 1974).

3.4 Shoot Look Shoot with Multiple Targets

When multiple targets are present, the range of alternative actions expands considerably. There are many possibilities for the nature of the feedback, objective function, and tactical constraints. Here we examine only feedback about target status. As in the case of a single target, the significance of this feedback is that it permits one to avoid shooting at targets that are already dead.

Our goal will be to kill as many targets as possible, or more generally as much target value as possible, on the average. It would of course be equivalent to minimize the value of the *surviving* targets, and it will sometimes be convenient to do that. This is a general-purpose goal – a more specific goal will be considered in Chapter 4. The general-purpose goal is appropriate when the benefit from killing targets is proportional to the number killed. An example of a situation where this is *not* true would be an attack on a redundant communications system consisting of three repeaters. Killing two of the repeaters is useless, since only one is required, so the value of doing so is not 2/3 of the value of killing all three.

3.4.1 Identical Shots and Identical Targets with a Time Constraint

We assume that the shots all have independent effects, with single shot kill probability $p = 1 - q$, and that the time constraint takes the form of a limitation on the number of salvos that are possible. The number of salvos might be limited because of the time to reload a launcher or because of the time required to assess target status – see Section 4.2.1 for a geometric analysis of one self-defense situation. Were it not for this constraint on the number of salvos, the optimal firing policy would be "shoot once at every live target, assess the status of each target, and then repeat until there are either no weapons left or no targets left." The only difficulty with that policy is that there might not be enough time to carry it out.

Consider an air defense system that has nine missiles available and has detected a flight of three enemy aircraft. Considering the speed of the aircraft and the time required to carry out each engagement, including status assessment, assume there is time for only two salvos. The question is, how many missiles should be fired against each of the aircraft in the first salvo? Launching three missiles against each aircraft would leave no missiles available to deal with any survivors of the first salvo and might waste several missiles on targets killed by other missiles. Launching only one missile against each aircraft might leave several aircraft still alive with only one salvo remaining. Might it be optimal to launch two missiles against each aircraft, or, for the sake of argument, two missiles against one aircraft and three against each of the other two? Any of these answers might be correct, depending on circumstances. What we need is a reliable method for determining the best firing policy.

Since all targets are identical, each salvo should be spread as evenly as possible over the remaining targets. The firing question thus reduces to determining how many shots x to spend in each salvo. Let random variable Y be the number of targets surviving out of t when x shots are fired, and let $F_n(s,t)$ be the number of targets that can be expected to survive n additional salvos when s shots and t targets remain. Statements such as the previous sentence are often the introduction to an analysis by dynamic programming, as is the case here. Since all targets will survive if there are 0 salvos remaining, we have $F_0(s,t) = t$, and for $n > 0$ we have the iteration

$$F_n(s,t) = \min_{0<x\leq s} E(F_{n-1}(s-x,Y)). \tag{3.8}$$

Equation (3.8) is at the heart of the dynamic programming procedure. It says that the average number of targets surviving is the same as the average number that survive when the number of salvos is reduced by 1, the number of shots is reduced by x, and the number of targets surviving the current salvo is Y. The expected value is needed because of the randomness of Y. Note the distinction between Y, the number of survivors of the salvo taken when n salvos remain, and $F_n(s,t)$, the (minimized) expected number of survivors of all n remaining salvos. The decision variable x in (3.8) can be any integer between 0 and s and should be chosen to minimize the expected number of targets surviving all remaining salvos. The desired probability distribution of Y is $Surv(y;x,t,q)$ as given by equation (2.31), since x shots should be spread as evenly as possible over t surviving targets.

The basic idea in solving (3.8) is to construct a sequence of tables indexed by n, starting at 0 and gradually increasing n until the number of salvos available is reached. Since $F_0(s,t)$ is known for all s and t, (3.8) permits $F_1(s,t)$ to be computed for s and t ranging from 0 up to the maximum number of shots and targets. Once $F_1(s,t)$ is known, (3.8) permits the computation of $F_2(s,t)$ over the same range of s and t, since everything on the right-hand side is now known. With $F_2(s,t)$ known, $F_3(s,t)$ can be computed, etc. Finding the minimum is done by simply trying all feasible values for x. As computations proceed, the optimal value for x is recorded and written in a separate table, for later recall when needed. One feature of this procedure is that, if the number of salvos is limited to (say) 3, then the optimal policies for 1 and 2 salvos must be determined before considering the optimal policy when three salvos remain. This is actually a tactical advantage, since a three-salvo problem will become a two-salvo problem right after the first salvo is fired.

It may seem that the procedure described above is simply optimization by exhaustion, since all possibilities for x are considered in the minimization step. This is not the case. The computations are more efficient than exhaustion because it is never necessary to consider firing policies that encompass multiple salvos, as might be thought necessary. We never have to consider complicated strategies such as "fire 5 in the first salvo, and then 3 in the second unless only one target remains, in which case … and then in the third salvo …." Each computation simply minimizes the number of survivors with one less salvo remaining, until the actual number of salvos is finally considered. Equation (3.8) is an application of dynamic programming (DP) to a Markov decision process (Puterman, 1994), a technique that is often useful in optimization problems that involve exhaustible resources and time. We will employ it repeatedly, generally referring to "stages," rather than salvos. DP is characterized by a functional equation like (3.8) where the objective function ($F()$, in our case) appears, albeit with different arguments, on both sides of the equation.

Correct identification of the state is a crucial part of DP problem formulation. It may help to imagine a "change of command" in the middle of the process, with the state of the process being whatever information the old commander should transfer to the new one. In the problem under consideration, the state is (n, s, t), where the three variables are the number of stages left, the number of shots left, and the number of targets still surviving. The past may have much more detail than that, but all such detail is irrelevant for purposes of future action – three numbers suffice. Once the state is identified, the DP procedure is always to define an objective function that depends only on the state variables, and then write down an equation such as (3.8) that has the effect of advancing the stage index by unity.

Sheet "DPShooter" of *Chapter3&4.xls* solves a firing problem with three salvos by this technique. The single-shot kill probability can depend on the salvo index, so there are three such inputs, rather than just one. It is suggested that the reader experiment with that sheet, especially if DP is a new subject. Do not forget to press the "DP Solve" button after changing the inputs, since results are not updated automatically. Sometimes the optimal results are obvious. For example, the dynamic program concludes that all remaining shots should be fired when there is only one salvo remaining, which should require no analysis whatever. Other results are not obvious, especially when the kill probabilities differ among the stages. To see the code that accomplishes the optimization, press ALT-F11 to get the VBA editor, and then double-click sheet "DPShooter" in the project explorer. The code first reads the numbers of survivors with no salvos left from the spreadsheet, and then optimizes shooting with 1, 2, and 3 salvos remaining, following (3.8), and outputting those results to the spreadsheet.

The exact computations described above can be approximated. The simplest of these approximations is to apply expected value analysis (EVA) as introduced in Chapter 1, combined with relaxing the integer requirement for the number of shots per target. Thus, if x shots are spread evenly over t targets, the survival probability of each target is $q^{x/t}$, so, using EVA, $tq^{x/t}$ targets survive for the next salvo. If t_i is the number of targets remaining after the ith salvo, then (3.8) can be replaced by $t_{i+1} = t_i q^{x_i/t_i}; i > 0$, with t_0 being the given initial number of targets. If there are n salvos, then t_n should be minimized subject to the constraint that $\sum_{i=1}^{n} x_i \le s$, where s is the number of shots available. Sheet "EVA" of *Chapter3&4.xls* employs Excel's Solver in this manner. This is a much simpler analysis than (3.8), but the results of the two methods agree only approximately. Exercise 10 is related.

3.4.2 Variety of Shots and Targets, No Time Constraint

In this case shots can be made one at a time, thereby avoiding the possibility of shooting at a dead target. The complication arises because it is not clear which weapon to assign to which target (in this section we refer to "weapons", rather than "shots", to prevent confusing statements such as "we first shoot shot 2"). Since the targets are all different, we adopt the goal of killing as much target value as possible, with target j having value v_j. Let p_{ij} be the known kill probability if weapon i is assigned to target j, possibly computed using one of the methods of Chapter 2. A recursive relationship such as equation (3.8) is still possible, but we can no longer merely keep track of the *numbers* of remaining weapons and targets, but must instead keep track of the *sets* of those. Call those sets S and T, and let $V(S,T)$ be the maximum amount of target value that can be killed, on the average, if the set of S weapons is optimally applied to the set of T targets. If either S or T is empty, then of course $V(S,T) = 0$.

$V(S,T)$ can be found by a dynamic programming recursion. Since each shot either kills its target or not, and since all shots are independent, we have

$$V(S,T) = \max_{i \in S, j \in T} \{p_{ij}(v_j + V(S-i, T-j)) + (1-p_{ij})V(S-i,T)\}. \tag{3.9}$$

If the first shot by weapon i is aimed at and kills target j, then the total value killed, on the average, is $v_j + V(S-i, T-j)$. If the first shot misses, then the total average value killed is $V(S-i,T)$. Averaging the two possibilities, we have (3.9). To use (3.9), we first evaluate $V(S,T)$ for all target subsets when S has only one weapon in it. Next we can solve all problems where S has only two weapons in it, since S–i will have one weapon if S has two, and so on. The number of evaluations of (3.9) required to ultimately complete this task can be very large, especially if S and T are large sets. Bellman (1961) refers to this as "the curse of dimensionality." Dynamic programming still works in principle, even in the presence of the curse. As a practical matter, however, it can only be used on small problems.

If $v_j = v$ and $p_{ij} = p$ for all i and j, as in Section 3.4.1, then (3.9) simplifies considerably because the sets S and T merely need to be counted. Let s and t be the numbers of weapons and targets, and let X be a binomial random variable with s trials and p success probability. X can be interpreted as the number of effective weapons. The number of kills will be X unless the marksman runs out of targets, so $V(S,T) = E(\min(X,t))$, a relatively simple computation. Anderson (1989) notes that the same formula applies as long as p_{ij} does not depend on j. See also Przemieniecki (1990) and Exercise 7. The same results could be obtained from (3.9) with a very large value of n, of course, but with much more difficulty. The associated optimal firing policy is trivial: "shoot at any live target as long as you have any weapons left."

Since (3.9) is problematic for large problems, we next develop some bounds on $V(S,T)$ in the general case.

A simple lower bound $V\_(S,T)$ can be constructed by computing the optimal pair $(i^*, j^*) = \arg\max_{i \in S, j \in T} v_j p_{ij}$. This is the "myopic" firing policy that maximizes only the immediate gain – every shot is taken as if it were the last. Equation (3.9) (with (i^*, j^*) substituted for (i, j) and $V\_()$ replacing $V()$ on both sides) must still be employed to evaluate $V\_(S,T)$, so exact evaluation is still difficult. Monte Carlo simulation would be a good way to approximate $V\_(S,T)$, but the myopic policy is trivial to implement in any case because knowledge of $V\_(S,T)$ is not needed to do so.

Example 2: The myopic policy is not always optimal. Consider two weapons and two targets, with $\mathbf{P} = \begin{bmatrix} 1 & 0.9 \\ 0.9 & 0 \end{bmatrix}$ and $\mathbf{v} = (1,1)$. $\mathbf{P}$ is a matrix of kill probabilities, with rows for weapons and columns for targets. The myopic policy will assign weapon 1 to target 1, since a gain of 1 is better than a gain of 0.9. Weapon 2 is then useless, since it has no capability on target 2. The optimal policy will first assign the second weapon to target 1, and then the first weapon to target 1 in case of failure, or to target 2 in case of success. The optimized total score is 0.9(1+0.9) + 0.1(0+1) = 1.81 – a substantial improvement over the myopic score of 1.0.

An upper bound applicable in more general circumstances is derived in the next section.

3.4.3 Variety of Shots and Targets, Unreliable Feedback

So far, we have only considered problems where the information feedback consists of a reliable live/dead report for each target. There are many other possibilities. We might measure things such as the target's radar cross section, its temperature or the kind of noise that it is making. Each of these is related to whether the target is live or dead, but not a precise statement about which is the case. Some of the relevant literature for problems of this sort will be mentioned in the next section, but the analyses tend to be beyond the scope of this book. In this section we consider only an upper bound on the effects of firing that is valid no matter what kind of information about target status is fed back, as long as the single shot kill probabilities against live targets remain constant. When information is imperfect, it is possible that shots will be made against targets that are already dead. As long as the target is still alive, however, the kill probability is assumed to remain p_{ij} as long as the shot is of type *i* and the target is of type *j*.

Fixed sets of weapons and targets are given. Define a collection of indicator random variables (random variables whose only possible values are 0 and 1) that can be associated with any firing policy:

$Y_{ij} = 1$ if target *j* is killed by weapon *i* (for each *j*, at most one of these is 1),
$Y_j = 1$ if target *j* is killed, and
$X_{ij} = 1$ if weapon *i* is assigned to target *j* (for each *i*, at most one of these is 1).

The firing policy will induce many dependencies between these random variables, so independence assumptions among them are not appropriate, but the collection is still useful for formulating the problem of finding the optimal policy. We first note that $Y_j = \sum_i Y_{ij}$, and that the total value killed is $Z = \sum_j v_j Y_j$. The problem (call it P1) of finding the optimal policy can therefore be posed as maximizing *z*, the expected value of *Z*, subject to the following constraints:

(a) $\sum_j X_{ij} \le 1$ for all *i* with certainty,

(b) $Y_j = \sum_i Y_{ij}$ for all *j* with certainty,

(c) $Y_j \le 1$ for all *j* with certainty, and

(d) other constraints.

The other constraints in P1 include the crucial relationship between X_{ij} and Y_{ij}, the essence of the firing policy. For example, the (probably foolish) policy of ignoring all information and simply making $X_{i1}=1$ for all *i* would probably kill target 1, but would also result in $Y_j = 0$ for $j > 1$. There are potentially an astronomical number of firing policies, since the decision about what to do next can depend in many ways on the information available. Nonetheless, regardless of the policy employed, we assume that $E(Y_{ij}) \le p_{ij} E(X_{ij})$, with strict inequality being possible because target *j* might already be dead when weapon *i* attacks it. Now, using lower case letters for expected values of random variables (so $y_j \equiv E(Y_j)$ and $x_{ij} \equiv E(X_{ij})$), we can construct a relaxation of P1 that we name P2:

maximize $z = \sum_j v_j y_j$ subject to

(a) $\sum_j x_{ij} \le 1$ for all i,

(b) $y_j \le \sum_i x_{ij} p_{ij}$ for all j,

(c) $y_j \le 1$ for all j, and

(d) all variables nonnegative.

P2 is a relaxation of P1 because a relationship that is true with certainty will also be true on the average, because sums and expected values can be interchanged, because $E(Y_{ij}) \le x_{ij}\, p_{ij}$ by assumption, and because the other constraints (d) of P1 have simply been omitted in P2, except for the part that requires non-negativity. Since P2 is a relaxation of P1, P2 provides an upper bound on what is achievable with any shoot-look-shoot policy. Since P2 is a linear program (Appendix B), it can be efficiently solved.

P2 has a direct interpretation where x_{ij} is the probability of using weapon i on target j, and y_j is the probability that target j is killed. In P2, (a) requires that weapon i not be used more than once, (b) requires that the effect of each weapon assigned to target j not exceed p_{ij}, and (c) requires that target j be killed at most once.

If b_i weapons of type i are actually identical, then constraints (a) of P2 can be changed to $\sum_j x_{ij} \le b_i$, a simple consequence of collecting terms with identical coefficients in P2. In that case the interpretation of x_{ij} is "average number of weapons of type i used on target j." The upper bound calculations are trivial compared to the solution of (3.9), especially if there are large groups of identical weapons.

The reader may wish to experiment with sheet "Optimal" of *Chapter3&4.xls*. The command buttons carry out three computations. One is the calculation of $V(S,T)$ using (3.9). This calculation is potentially time consuming, so try a small problem first. The second is the myopic calculation of Section 3.4.2, and the third is the upper bound just derived. The upper bound computations are carried out using data on sheet "LP_Upper," but that sheet need not be examined unless the reader is interested in the details of solving a linear program using Excel's Solver. The reader can verify that the upper bound for the small example introduced in Section 3.4.2 is actually exact, and that the upper bound is usually close to the truth for problems not designed to make it look bad.

Example 3: This example is designed to make the upper bound look bad. Consider a problem with one target and two weapons, each of which will kill the target with probability 0.5. The only available allocation of weapons produces a kill probability of 0.75, while the solution of P2 is that z can be made 1 by setting $x_{11} = 2$ and $y_1 = 1$.

Example 4. Suppose there are four hard targets with value 50 each, four soft targets with value 100 each, and five false targets with value 0. Each of these targets is either live or dead at all times, and some may be dead initially. The marksman has available five bombs and four missiles. A bomb will not kill a hard target, but will kill any other target with probability 0.5. A missile will kill a hard target with probability 1, or any other target with probability 0.5. Before each salvo, the marksman has an information system that examines the battlefield and possibly gives him a report about the existence and status of each target. Some targets may be omitted from the report, the report may be error prone, and reported information may be only indirectly relevant (a target's temperature, for example). The marksman may be initially ignorant of the nature of the battlefield, including the information specified above. Determination of an optimal firing policy would require much more detail about the marksman's initial state of uncertainty and the nature of the battlefield reports, but there is enough information given above to compute an upper bound on the results of using that policy. All that is required for computing the upper bound are the sets of weapons and targets, the target values, and the kill probability matrix. Although the false targets might be a significant problem to an actual marksman, the upper bound calculations simply omit them (see Exercise 11).

The upper bound can be easily computed even in circumstances where finding the optimal policy according to (3.9) or (3.8) is computationally impossible. Linear program P2 is solvable in a few seconds even for thousands of weapons and targets.

3.5 Further Reading

The optimal use of information is a difficult subject in general, and firing problems are no exception. This is a bit of an analytic crisis for military analysis, since modern military systems rely heavily on the timely acquisition and distribution of information. Information is not free, so there is sometimes a need to evaluate its effectiveness in quantitative terms, but evaluation is hampered by the analytic intractability of many of the associated problems. The models presented in this chapter, while hardly trivial, are only a small start on a large problem.

Error-prone target status information is common in reality, but not discussed above except for the bound in Section 3.4.3. Such information leaves the marksman in doubt about the status of targets, so it would be natural to introduce a probability distribution for the state of the target or the states of each of several targets. One can apply Bayes theorem to update the target state distribution(s) after each report. In some cases a myopic or "greedy" strategy something like "always shoot at the target most likely to be alive" can be shown to be optimal (Manor and Kress (1997), Glazebrook and Washburn (2004, Section 1)). Aviv and Kress (1997) discuss some simple firing procedures that are nearly optimal. If targets differ, but there is only one kind of weapon, the shooting process may be indexable in the sense that one should always shoot at the target with the highest numerical index (Glazebrook and Washburn, 2004, Section 4). More generally, this line of thought ends up with a difficult partially observable Markov decision process where the state of the process is itself a probability distribution (Yost and Washburn (2000), Glazebrook and Washburn (2004, Section 2)). Since the mere application of Bayes theorem can itself be data-intensive, and since Bellman's curse is operating, it is fair to say that techniques such as these are (in 2009) some ways from practical implementation. This situation might change as computers get more and more capable. In the meantime, lacking theoretical guidance about optimal firing policies, military judgment about dealing with uncertainty is often built into combat simulations instead.

A practical marksman can be in doubt about more than the target's functional status. The very identity of the target can be in doubt, witness the occasional instance of military fratricide. False alarms can happen, particularly in antisubmarine warfare where there are many phenomena that can lead to the appearance of phantom submarines. There are very few quantitative analyses of how to shoot optimally in such circumstances.

Exercises

(1) Suppose there are four shot types, with the cost and kill probability vectors being **c** = (1,2,3,4) and **p** = (0.2, 0.3, 0.4, 0.5).
(a) If there is one shot of each type, in what order should the shots be taken, what is the resulting kill probability, and what is the average cost of shooting?
(b) Same as part (a), except that there are unlimited numbers of each type.
Ans: In part (a), the shots should be taken in order 1, 2, 3, 4, the kill probability is 0.832, and the average cost of shooting is 5.624. In part (b), only shot type 1 should be used, the kill probability is 1, and the average cost of shooting is 5.

(2) Suppose you are to make eight shots at a target, each of which has a circular normal dispersion error with standard deviation 2 and a cookie-cutter lethal radius R. Which of the following three situations will give the highest kill probability?
(a) $R = 1$, and there is no bias error.
(b) $R = 2$, there is a circular normal bias error with standard deviation 5, and an observer reports miss distances as in the SAS procedure.
(c) $R = 3$, there is a circular normal bias error with standard deviation 5, and there are no observer reports
Ans: The first two kill probabilities are $1-\exp(-1)$ and $1-\exp(-2.64)$, so (b) is much better than (a). The answer for (b) uses (3.7) and has nothing to do with the number 5, since the SAS procedure eliminates the bias error. The exact P_K for (c) cannot be given without examining patterns of 10 shots as in Chapter 2, but an upper bound on the kill probability is $1-\exp(-0.64)$, so it is surely the worst of the three situations. The best of the three situations is thus (b), and the kill probability is nearly 1.

(3) With the three miss probabilities all set to 0.5, sheet "DPShooter" of *Chapter3&4.xls* claims that the optimal number of shots to fire with 2 salvos, 12 shots, and 5 targets remaining is 6. This is one of a small number of cases where the number of shots fired is not a multiple of the number of targets – one target is shot at twice, while the other four are shot at only once. The sheet also claims that shooting six shots, plus optimal firing in the last salvo, will result in 0.4993 targets surviving out of 5. Verify that the same advice is given with three salvos remaining if the kill probability for the final salvo is set to 0. Do not forget to press the "DP Solve" button, since spreadsheet output will not be updated unless you do. Setting the final kill probability to 0 should have the same effect as having one less salvo, so this is an exercise in verification.

(4) In the same circumstances as Exercise 3, Marie must decide how many shots out of 12 to fire at 7 targets when there are 3 salvos left. The "DPShooter" sheet advises firing 7 shots, but, for one reason or another, she only fires 4. The result of this is that she kills 2 targets, so she has 8 shots and 5 targets remaining when there are 2 salvos left. If she fires optimally in the last 2 salvos, how many targets will she kill, on the average, in the last 2 salvos?
Ans: 1.2227 will survive, so she will kill 3.7773, on the average. Given that there are 8 shots, 5 targets, and 2 salvos remaining, the results do not depend on how that state happened to arise. This is the fundamental fact on which the whole computational scheme is based.

(5) In the same circumstances as Exercise 3, the "DPShooter" sheet claims that the optimal number of shots is 2 when there are 4 shots, 2 targets, and 3 salvos remaining. If this advice is followed, the number of survivors with two salvos remaining will be either 0, 1, or 2. Compute the probabilities of those events and show by averaging the appropriate stage 2 results that the average number of survivors after all salvos are complete is indeed 0.375, as the sheet claims.
Ans: $(0.25)(0)+(0.5)(0.25)+(0.25)(1)=0.375$.

(6) Suppose there are three weapons and three targets of equal value, with the kill probability matrix being $\mathbf{P}=\begin{bmatrix}0.1 & 0.2 & 0.3\\0.4 & 0.5 & 0.6\\0.7 & 0.8 & 0.9\end{bmatrix}$. Use sheet "Optimal" to show that the maximized average number of targets killed with SLS is 1.683 with perfect feedback after each shot (the sheet also should show an upper bound of 1.711). The firing policy for achieving 1.683 has been computed but not displayed, since displaying the method in a spreadsheet would be too complicated. Static policies that do not utilize information feedback, as considered in Section 2.4.3, are much simpler to explain and evaluate. For example, the policy of firing weapon i at target i for all three targets would achieve an average number killed of $0.1 + 0.5 + 0.9 = 1.5$ or the policy of firing all three weapons at target three would achieve $1-(1-0.1)(1-0.6)(1-0.9)=0.964$. What is the best static policy that you can find? The average number of targets killed cannot exceed the SLS value of 1.683. Does it exceed the myopic value that is also given on sheet "Optimal?"
Ans: There are many ways of achieving it, but the best value achievable with a static policy is 1.5.

(7) Section 3.4.2 includes the formula $V(S,T)=E(\min(X,t))$ for the expected number of targets killed out of t when X is a binomial random variable with s trials and p success probability. Write a computer program that will take the

three inputs (s,t,p) and compute the average number of targets killed. If you are using Excel™, make use of the BINOMDIST() function.

Ans: If $(s, t, p) = (15, 10, 0.8)$, then the average number of targets killed should be 9.916. Even though the average number of killing shots $(s \times p = 12)$ is greater than 10, the average number of targets killed is slightly less than 10.

(8) Sheet "3CAL" of *Chapter3&4.xls* is a Monte Carlo simulation where SAS and 3CAL are compared as to their ability to hit a target in one dimension. Study the structure of that sheet to verify that the comparison is a fair one.
(a) Remarkably, both procedures seem to produce exactly the same results on the fourth shot, as can be seen by pressing F9 repeatedly. Explain analytically why the two aim points are always identical on the fourth shot.
(b) Use "SimSheet" of *AppendixC.xls* (see Appendix C) to confirm that SAS is superior to 3CAL on the tenth shot by showing that it has a higher hit probability.

(9) The miss distances in the SAS procedure are defined by (3.4) as a function of the dispersion errors, and the claim is made in that section that they are all independent of each other. The truth of this is not entirely obvious, since each miss distance depends on the same sequence of dispersion errors. Prove that X_2 is in fact independent of X_4. Hint: Since the miss distances are all normally distributed, they are independent if and only if they are uncorrelated, so it suffices to show that $E(X_2 X_4) = 0$.

Ans: Using (3.4), the product $X_2 X_4$ can be written out term by term. Terms where the factors have different subscripts can be ignored, since they average 0. Except for such terms, the product is $(E_1^2 - E_2^2)/3$, which also averages 0 because the dispersion errors are all identically distributed.

(10) Sheet "EVA" of *Chapter3&4.xls* solves a problem where 12 shots are allocated to 7 targets in 3 salvos, just like sheet "DPShooter." Set the miss probability in each of the three salvos to 0.5 and compare the number of survivors in the two models.
Ans: There are 1.295 survivors without EVA or 0.876 with EVA.

(11) Use sheet "Optimal" of *Chapter3&4.xls* to find the upper bound on target value killed for the problem specified in Exercise 4, as well as the myopic solution of a firing problem where information feedback is of the SLS type. This will require inputting a 10×8 matrix, with the last row being the values of the eight targets (false targets need not be input, since the allocations to them should clearly be 0). DO NOT press the "Compute Optimal" button, lest you hang up your computer on this large problem. Instead, press the "Compute Myopic and Upper" button to find the objective function in the two cases.

Ans: The myopic value is 446.88 and the upper bound is 450. Although the optimal firing policy for the shoot-look-shoot case is still unknown, the results are tightly constrained on both ends. The upper bound applies regardless of the type of feedback.

Chapter 4
Target Defense

Circle the wagons!

John Wayne

4.1 Introduction

In this chapter we consider problems where multiple attackers are approaching a target, hoping to kill it by overwhelming its defenses. The defense is armed with some anti-attacker weapons called interceptors, each of which can kill only the attacker to which it is assigned. The goal is to use the interceptors to maximize the survival probability of the target. Instead of assigning weapons to nondescript targets, as in Chapters 2 and 3, the defense now assigns them to attackers. The "targets" are now the things being defended.

One can also imagine situations where a target is to be defended by setting up screens or barriers that are not exhausted by a series of attacks. A convoy of ships might have a screen of destroyers designed to protect it from submarines, for example, or an important political leader might have a screen of agents designed to protect him from attack. Although such problems are well described as "target defense," they are analytically similar to Search theory problems, for which the reader is referred to Chapters 7 and 10. Problems in this chapter are characterized by having some kind of a resource that is gradually exhausted as attackers are engaged.

Most of the methods outlined below have been developed for situations where the defenders are anti-ballistic missiles (ABMs) and the attackers are either aircraft or missiles, particularly ICBMs, but there are also applications in other circumstances. Navy ships, for example, are particularly vulnerable to torpedoes, and anti-torpedo-torpedoes (ATTs) have been proposed as a countermeasure. A ship trying to defend itself against a group of incoming torpedoes using ATTs faces a problem that is analytically similar to a missile silo or city trying to defend itself using ABMs. In either case, attackers and defenders (interceptors) are expensive, individually assigned objects in short supply.

A complicating feature is that some attackers may attack the defense, rather than the targets directly. It is typical of air-to-ground campaigns, in fact, that the first step is usually to attempt to get control of the air by attacking the enemy's air force. The associated analytic problems can become quite complex from the viewpoint of either the attacker or the defender and will not be covered here. In this chapter, the target is attacked directly.

A. Washburn, M. Kress, *Combat Modeling*, International Series in Operations Research & Management Science 134, DOI 10.1007/978-1-4419-0790-5_4,

We will summarize basic results, emphasizing the important case where the attack size is unknown. Section 4.2 deals with a single target, section 4.3 deals with multiple targets, and Section 4.4 (Further Reading) discusses other results available in the literature.

4.2 Defense of One Target Against Several Identical Attackers

In this section there is only a single target, so the defender does not have the problem of distributing his interceptors over multiple targets. However, since multiple interceptors can be assigned to any given attacker, there remains the question of how many interceptors to assign to each sequential attacker. Much depends on how much is known about the attack size.

4.2.1 Known Attack Size

Assume that each of t attackers will independently kill its target with probability $p = 1 - q$ if not intercepted, and that the defender has s interceptors, each of which will independently kill an attacker with probability ρ, and all of which are to be used against the t attackers.

We first consider the situation where all s interceptors must be committed in one salvo. This is the same situation considered earlier in Section 3.4.1, except that the defender's goal is now to maximize the probability that the single target survives. As in Section 3.4.1, the defender should distribute the interceptors as evenly as possible over the attackers (Soland, 1987). Let k be the integer part of s/t, and let r be the remainder:

$$s = kt + r, \text{ where } 0 \le r < t\,. \tag{4.1}$$

When the interceptors are distributed as evenly as possible, $(t-r)$ attackers are assigned k interceptors, r are assigned $k+1$, and the probability that the target survives is

$$Q(s,t) \equiv \left[1-p(1-\rho)^k\right]^{t-r}\left[1-p(1-\rho)^{k+1}\right]^r. \tag{4.2}$$

Example 1: Suppose $p = 0.8$, $\rho = 0.5$, $s = 7$, and $t = 3$. Then $k = 2$ and $r = 1$. Each of the 2 attackers that are assigned 2 interceptors will kill the target with probability $p(1-\rho) = 0.2$, and the target will therefore survive both attackers with probability $0.8^2 = 0.64$, which is the first [ ] factor in (4.2). The second is $(1-0.8(0.5)^3)^1 = 0.9$, so $Q(7,3) = 0.576$.

The target survival probability $Q(s, t)$ can be approximated by permitting non-integer allocations of interceptors, s/t to each attacker:

$$Q(s, t) \cong (1 - p(1 - \rho)^{s/t})^t. \tag{4.3}$$

Equation (4.3) approximates $Q(7, 3)$ in Example 1 by 0.595; (4.3) will in all cases be at least as large as (4.2).

If sufficient time is available, shoot-look-shoot policies of the type considered in Section 3.4.1 may be possible. The same dynamic programming method used in that section will also work here, but with a different objective function. Let $F_n(s,t)$ be the maximum target survival probability when n stages, s interceptors, and t attackers remain. Since the goal is now maximization, equation (3.8) is replaced by

$$F_n(s,t) = \max_{0<x\leq s} E(F_{n-1}(s-x, Y));\ n > 0\,. \tag{4.4}$$

With 0 stages left, the number of interceptors remaining is irrelevant because there is no further opportunity to use them, and the target is reduced to hoping that all t remaining attackers fail in their mission. The probability of this is $F_0(s,t) = 1 - q^t$. If n is greater than 0, then the assignment of x interceptors to the current salvo will leave Y attackers, where Y is a random variable with the same properties as in Section 3.4.1. As in Section 3.4.1, the solution strategy is to use $F_n(s,t)$ to compute $F_{n+1}(s,t)$ for all relevant (s,t) pairs, until the stage index n is advanced to the number of interceptors initially available. An implementation for three stages is on sheet "DPDefender" of *Chapter3&4.xls.* Except for one additional input (the attacker kill probability p), sheet "DPDefender" looks and functions just like the sheet "DPShooter" that was employed in Section 3.4.1. See Soland (1987) for a more detailed derivation of (4.4).

The number of stages or salvos is sometimes derivable from geometric considerations. One such circumstance might be called "self-defense," where attackers are assumed to proceed at constant speed U directly toward their target, the marksman. The marksman launches interceptors with constant speed V at the attackers. After each engagement, it takes a certain amount of time to assess which attackers still survive. During that time, the surviving attackers move toward the marksman by a distance Δ, the product of U and the assessment time. If a salvo of interceptors is launched at range R, then the intercept will be at time $R/(U+V)$, since the relative speed is $U+V$, at which time the interceptors (and therefore the attackers) are at range $RV/(U+V)$. Let α be the ratio $V/(U+V)$. Allowing for attacker movement during the assessment time, we can say that that the next salvo of interceptors can be launched at range $R' = \alpha R - \Delta$, provided that that range is large enough to permit launch. See Figure 1. This argument can be repeated to find the launch range following R', and so on. If R_i is the ith launch range, then R_1 is determined by either the maximum range of the marksman's sensor or the maximum interceptor range, and succeeding ranges are given by

$$R_{n+1} = \alpha^n R_1 - \Delta \frac{1-\alpha}{1-\alpha^n}; n \geq 1, \tag{4.5}$$

as can be shown by induction (Exercise 3). As soon as R_{n+1} is too small to permit launch of a salvo of interceptors, n is the number of stages to be used in (4.4).

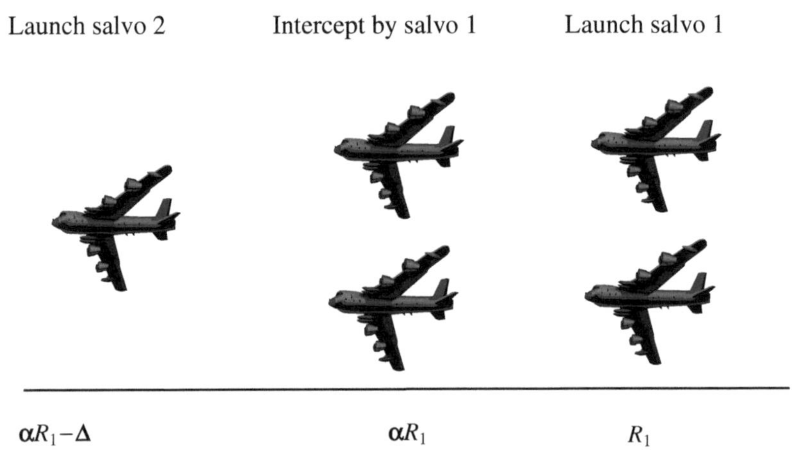

Figure 1: There are 2 attackers when salvo 1 is launched, and 1 attacker when salvo 2 is launched.

Example 2: The maximum range of a radar is 6000 m. A battery of anti-aircraft missiles is available to defend a ship against attacking surface-to-surface (SSM) missiles that approach at 100 m/s. The defending missile speed is 200 m/s. It takes 2 s to determine the outcome of any engagement, and any SSM that gets within 200 m of the ship will detonate. How many salvos are possible without risking a detonation? To answer, use $R_1 = 6000$ m, $\alpha = 2/3$ and $\Delta = 200$ m to calculate the sequence of launch ranges in meters: 6000, 3800, 2333, 1356, 704, 269, –21. Although the sixth salvo can be launched before any SSM detonates, the intercept will not occur until range $\alpha R_6 = 179$ m, which is smaller than 200 m, so only five salvos are actually possible. Note that the salvos will not be uniformly spaced in time, since the initial engagements take a long time to complete.

The analysis leading to (4.5) is more general than that in Wagner et al. (1999, ch. 13) or Przemieniecki (2000, Section 6.5) in allowing for an assessment advance Δ, but less general in that those two references consider geometric situations where the attackers do not proceed directly toward the marksman; i.e., situations that are not necessarily self-defense. In such situations see Section 3.4.1, since "survival probability" is no longer the appropriate measure of effectiveness. If the attackers are aircraft, it may also be possible to shoot at them on the way out, as well as on the way in.

Suppose now that the attackers arrive one by one, widely separated in time, and moving so fast that only one salvo is possible for each attacker. Let m_i be the number of interceptors allocated to the ith attacker; i=1,…,n in an optimized firing schedule. We have argued above in justifying (4.2) that the m_i should be as equal as possible (a "flat" defense), but the reader may have intuitive feelings that a "tapered" defense where early attackers see more interceptors would be more desirable. Intuitively, it would be regrettable if there were several unused interceptors remaining after the target is killed, since the interceptors have no other use than to defend the target. We might therefore expect early attackers to be hit hardest in the optimized schedule. In spite of intuition, it remains true that the defense should be flat as long as the attack size is known.

A tapered defense can be useful when the attack size is *unknown*. Sections 4.2.2, 4.2.3, and 4.2.4 deal with three versions of the problem where n is unknown, each of which leads to a tapered defense of some kind.

4.2.2 Bayesian Defense

The attackers are assumed to show up one at a time, forcing the defense to guess the number of successors (if any) on each occasion. Much will depend on the defense's ability to anticipate the unknown ultimate size of the attack, N.

The defense has a stockpile of m interceptors, each of which has kill probability ρ. These interceptors are available for the defense of a single target against a sequence of attackers, each of which has kill probability p if not intercepted. It is assumed that N does not exceed some known number n, and that the probability law for N is known (hence the term "Bayesian," which implies the existence of a prior distribution). Let m_i be the number of inceptors allocated to the ith attacker. The objective is to determine the firing schedule m_1, …, m_n that maximizes the probability of surviving all N attackers, subject to the constraint that $m_1 + \cdots + m_n = m$. The shoot-look-shoot analysis of Section 4.2.1 would apply if N were known, with N stages because the attackers arrive one at a time. However, random variable N is unknown, except for its probability law.

We will solve the problem using dynamic programming. The state of the process is (s, i), where s is the number of interceptors remaining and i is the number of attackers that have already arrived and failed to kill the target. The decision required is to determine the number of interceptors x to use against the next attacker. The i part of the state is important because if $i = n - 1$, for example, then it is clear that s is the best choice for x, whereas it might otherwise be wise to reserve some interceptors against the possibility of additional attackers. The objective function is $F(s,i)$, the maximum survival probability of the defended target.

To develop a recursive formula for $F(s,i)$, we must first recognize that there may not be any future attackers at all, in which case survival is certain. The probability that there will be at least one more attacker, given that i attackers have already arrived, is $Q_i \equiv P(N > i \mid N \geq i)$. Since the probability law for N is known, so are these "continuation probabilities" Q_i. If there is at least one more attacker, which happens with probability Q_i, and if the next attacker does not destroy the target, then the next state will be $(s-x, i+1)$. The desired recursion is thus

$$F(s,i) = 1 - Q_i + Q_i \max_{0 \leq x \leq s} \left\{ \left[1 - p(1-\rho)^x \right] F(s-x, i+1) \right\}. \tag{4.6}$$

The expression in [ ] is the probability that the next attacker does not destroy the target.

It is clear that $F(s,n) = 1$ for all s, since survival is certain if all attackers have already arrived and failed to kill the target. Equation (4.6) can therefore be used to compute $F(s,n-1)$ for all $s \leq m$, then $F(s,n-2)$ for all $s \leq m$, etc., until finally $F(s,0)$ is obtained. $F(m,0)$ is then the best overall survival probability. In the process of doing the computations, the optimal allocation of interceptors can be recorded as the optimal firing schedule. See Exercise 6.

4.2.3 The Maximum Cost Defense Against a Shoot-Look-Shoot Attack

We assume here, as in Section 4.2.2, that the number of attackers is unknown, and that a firing schedule for the defensive interceptors must be set up for use as attackers arrive one at a time. However, no probability distribution is given for the total number of attackers. Instead, the defense takes the point of view that any target defended by a finite stockpile of interceptors can be killed if sufficiently many attackers are committed, and that the proper goal is therefore to maximize the cost (measured in attackers) of killing the target. If this number turns out to be so large that the attack does not take place, then so much the better, but in any case the defensive goal is to make the target as hard to kill as possible. The same defensive tactic will also delay any possible kill, thus allowing time for other defensive measures. The attacker is assumed to have a shoot-look-shoot capability, so there will be no wasted attackers. The defense's goal is to make sure there are no cheap victories.

The objective of maximizing the average number of attackers required to kill the target can be accomplished by once again using dynamic programming, but the state description is simpler than in Section 4.2.2. The state of the process is simply the number of interceptors m remaining, and the objective function is $c(m) \equiv$ "the average number of additional attackers required to kill the target if m interceptors remain." Suppose x interceptors are allocated to the next attacker. The probability that the next attacker kills the target is then $p(1-\rho)^x$, where p and ρ

are the kill probabilities of attackers and interceptors, respectively. If the next attacker fails to kill the target, then the next state will be $m - x$. Therefore, since at least one attacker is required in any case,

$$c(m) = 1 + \max_{0 \le x \le m} \left\{ \left(1 - p(1-\rho)^x\right) c(m-x) \right\}. \tag{4.7}$$

If $m = 0$, (4.6) requires $c(0) = 1 + (1-p)c(0)$, which has the solution $c(0) = 1/p$. This is the average number of attackers required to kill an undefended target. For $m > 0$, the option $x = 0$ can safely be ignored, since at least one interceptor should be used in any case. Equation (4.7) can therefore be used to determine $c(1)$, then $c(2)$, etc., recording the maximizing value of x at each stage (call it $x^*(m)$). Sheet "MaxCost" of *Chapter3&4.xls* is an implementation.

Example 3: Suppose $p = 0.8$ and $\rho = 0.5$. Then $c(0) = 1.25$, and (the maximizing element is underlined)

$c(1) = 1 + 0.6c(0) = 1.75$, and $x^*(1) = 1$
$c(2) = 1 + \max\{\underline{0.6c(1)}, 0.8c(0)\} = 2.05$, and $x^*(2) = 1$
$c(3) = 1 + \max\{0.6c(2), \underline{0.8c(1)}, 0.9c(0)\} = 2.40$, and $x^*(3) = 2$

Continuing in this manner, we find that $c(m) = 1.25, 1.75, 2.05, 2.40, 2.64, 2.92, 3.16, 3.38, 3.63, 3.84, 4.04, 4.27$ for $m = 0, 1, \ldots, 11$, and also $x^*(m) = 0, 1, 1, 2, 2, 2, 3, 3, 3, 3, 3, 3$. If 11 interceptors remain, three should be used against the first attacker, then $x^*(8) = 3$ should be used against the second, $x^*(5) = 2$ against the third, $x^*(3) = 2$ against the fourth, and $x^*(1) = 1$ against the fifth. The sixth and subsequent attackers would not be opposed, since the defenses are exhausted.

The function $c(m)$ is not tactically necessary, even though it is the focus of optimization – it suffices to remember the function $x^*(m)$. One might, however, use $c(m)$ as a measure of effectiveness for making an interceptor quantity versus quality decision (see Exercise 7).

The philosophy of conducting the defense to maximize the cost of killing a target sees occasional use in more general combat models. Ravid (1989) applies it to arranging anti-aircraft defenses, as outlined in the next example.

Example 4. Suppose there are two methods of defending an airfield. The "Near" defense arranges the defending guns so that they are closely spaced and near the airfield. Any attacking aircraft will have to pass near at least one of the guns, so the chances of shooting down the aircraft are relatively high (0.2). The trouble with this defense is that it acts after the attack, so the bombs will strike the airfield even if the aircraft is shot down. The "Far" defense arranges the guns far enough away that any attacking aircraft must survive in order to drop its bombs. The trouble with this defense is that the chances of shooting down the aircraft are relatively low (0.1). The problem is to choose one of the two defense types.

The Far defense would be optimal in a situation where the fate of the aircraft is irrelevant, and the only object is to protect the airfield against the next attack.

However, the usual context of attacks on airfields is that any airfield can be attacked if the attacker is willing to pay the price in terms of aircraft lost, in which case the defense should have the goal of making successful attacks as expensive as possible. The cost of a successful attack against the Near defense is 0.2 aircraft, on the average. Over its lifetime, an aircraft against the Far defense will make an average of 1/0.1 attacks (the mean of a geometric random variable), but the last one will not be successful, so there are only nine successful attacks per aircraft lost, and the cost of a successful attack is therefore 1/9. The best defense is revealed to be the Near defense, even though it stands no chance of preventing the immediate attack.

4.2.4 Prim–Read Defense

The assumptions in this section are the same as in Section 4.2.3, except that the attackers are no longer assumed to have a shoot-look-shoot capability. The attackers still arrive sequentially, but a certain number (say n) out of a large stockpile must be irrevocably committed to the target. Let $p(n)$ be the probability that the target is killed by one of n attackers, and let $\lambda = \max_{n>1} p(n)/n$. λ is the largest possible kill probability per attacker. The objective of a Prim–Read defense is to make λ as small as possible with whatever interceptor stockpile is available to the defense, the idea being to prevent cheap kills. The idea was first proposed by Prim and Read as a method for defending targets against ICBM attack (Read, 1958; Burr et al., 1985).

The problem of minimizing the defensive stockpile required to achieve a given λ turns out to be much easier analytically than the problem of minimizing λ for a given stockpile, so much so that a problem of the latter type is most easily solved by guessing values for λ until the calculated stockpile is whatever happens to be available. This technique is illustrated below.

Let m_i be the number of interceptors allocated to the ith attacker. Making the usual independence assumptions, and letting p and ρ be the kill probabilities of attackers and interceptors,

$$p(n) = 1 - \prod_{i=1}^{n} \left(1 - p(1-\rho)^{m_i}\right); \quad n \geq 1, \tag{4.8}$$

and the central problem is to minimize $\sum_{i=1}^{\infty} m_i$ subject to the constraints that $p(n) \leq \lambda n$ for all $n \geq 1$.

Suppose, for example, that $p = 0.8$, $\rho = 0.5$, and that there are $m = 11$ interceptors available. Our initial guess might be that 11 interceptors should be sufficient to guarantee that the kill probability per attacker will not exceed (say)

$\lambda = 0.3$. We now consider the problem of minimizing the number of interceptors required to guarantee that the maximum kill probability per attacker does not exceed 0.3, hoping that the answer is 11. From (4.8), $p(1) = p(1-\rho)^{m_1}$. Since $p(1)$ must not exceed 0.3, the smallest possible value for m_1 is 2, so we take $m_1 = 2$. We therefore have $p(2) = 1 - 0.8(1 - p(1-\rho)^{m_2})$. The smallest value of m_2 for which $p(2) < 0.6$ is 1, so we take $m_2 = 1$. We therefore have $p(3) = 1 - (0.8)(0.6)(1 - p(1-\rho)^{m_3})$. The smallest value of m_3 for which $p(3) < 0.9$ is 1 ($p(3) = 0.904$ when $m_3 = 0$, which is just barely too large), so we take $m_3 = 1$. We can take $m_i = 0$ for $i \geq 4$, since the return per attacker cannot exceed 0.25 if there are four attackers. The total number of interceptors required to guarantee that the kill probability per attacker does not exceed 0.3 is therefore 2 + 1 + 1 = 4. Eleven interceptors are evidently sufficient for a smaller value of λ than the assumed 0.3. The next step is to guess a smaller value (see Exercise 9) and repeat the above calculations. The calculations are easy because the product in (4.8) can be formed sequentially, with the first $(n-1)$ factors being known when m_n is being determined. See sheet "PrimRead" of *Chapter3&4.xls* for a spreadsheet implementation. The simplicity of the calculations makes up for the fact that they must typically be repeated several times.

Although a Prim–Read defense can certainly be constructed for a single target, the technique is more naturally applied to a group of targets, using the same value of λ for every target in the group. If the targets differ in value from each other, one simply introduces a target value into (4.8), and λ becomes "maximum value killed per attacker."

An implicit assumption in setting up a Prim–Read defense is that the attacker can determine the defensive firing schedule before making his own allocations. There may be good physical reasons for assuming this, as in the ABM problem originally considered by Read. However, it may also be true that the attacker has just as much trouble ascertaining defensive allocations as vice versa. In the latter case, a Prim–Read defense is probably a mistake. The Prim–Read defense of several identical targets would treat all targets equally, for example, whereas the best defense may be to abandon half of the targets in order to construct a strong defense of the remainder, as in Blotto games (Section 6.2.3).

4.3 Defense of Multiple Targets Against ICBM Attack

The problem of defending a collection of targets from ICBM attack has been much studied. There was already enough literature in 1970 to prompt a review (Matlin, 1970), and research has continued in the ensuing years as command and control improvements have permitted increasingly ambitious planning tools. The last three chapters of Eckler and Burr (1972) are devoted to the topic and are a good summary of work up until that time. The more recent book by Przemieniecki (2000) includes relevant chapters titled "Strategic Defense" and "Theater Missile

Defense." Here we review the basic facts of ICBM technology, and the possible applicability of the models introduced earlier.

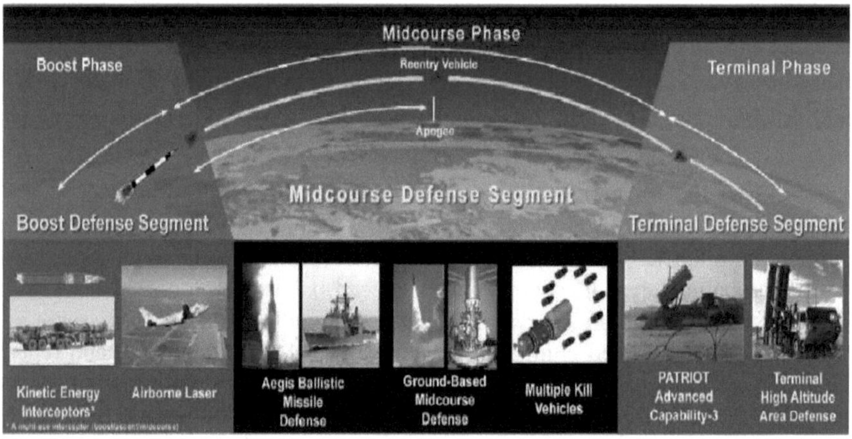

Figure 2: United States ballistic missile defense systems current in 2007 (Source: Ballistic Missile Defense Organization).

The active life of an ICBM is generally supposed to have three phases, as illustrated in Figure 2:

1. Boost phase. The ICBM is vulnerable during this phase because it is emitting lots of heat, and a choice target because killing it will also kill all of the re-entry vehicles that it carries. Unfortunately, boost phase occurs with little warning and does not last long. Kills are therefore hard to achieve.
2. Mid-course phase. This phase is outside the atmosphere, which means that light but effective decoys can be deployed. Most of the ICBM's active time is spent in this phase. The Brilliant Pebbles part of the Strategic Defense Initiative (SDI) functions here.
3. Terminal phase. Once the ICBM payload re-enters the atmosphere, any light decoys burn up or slow down, exposing the heavier re-entry vehicles. Terminal phase is brief, but the compensation for the defense is that the targets are clear and the distances to be traveled are relatively short. Defense is usually conducted with relatively small, fast missiles such as those launched by Patriot.

The three ICBM phases may last on the order of half an hour in total. Defensive systems are also sometimes arranged in phases or "layers," with separate defensive systems for each phase.

It is nearly always assumed that each side is aware of the total forces available to the other. If the defender knows in addition how the attackers are distributed over the targets, then his defense of each target might follow the lines of Section 4.2.1.

Kooharian et al. (1969) examine a generalization where the targets differ from each other.

If the defender does not know the distribution of attackers over targets, then the situation can quickly become complicated. Sections 4.2.2 through 4.2.4 are each different methods of defending individual targets without having specific information about how many attackers to expect.

The Prim–Read defense can be viewed as a conservative reaction to a situation where the defenses must be arranged in a manner visible to the attacker, as may be true of terminal interceptors that must be physically located near their defended targets. The analysis of Section 4.2.4 generalizes easily to the defense of several targets, even if they have different values, as long as the attackers are all identical. Washburn (2005) describes a kind of Prim–Read defense for a generalization where the attackers are indistinguishable to the defense, but nonetheless of mixed types, including decoys.

The attacker may have the option of extending the attack in time, in which case the defender may be confronted with situations where it is feasible to destroy a given attacker, but not necessarily wise to do so. For example, suppose there are 10 targets, 20 attackers, and a defense with 10 perfect interceptors and the capability of determining the target to which each attacker is committed. The defense might employ the principle that no target will be killed as long as the capability to intercept attackers remains, but doing so may not be wise. If that principle is employed, the attacker can engage all 10 targets by first exhausting the defenders and then aiming one attacker at each of the undefended targets. The defender could save several targets by defending only a subset, ignoring any attackers devoted to targets outside the subset (the Blotto games of Section 6.2.3 are a formalization of this idea). The superiority of this tactic is sometimes recognized when the targets are things like ICBM silos, but it is hard to imagine not defending a specific city because it does not happen to be in some arbitrary subset. Analyses sometimes assume implicitly that the principle will be employed. Miercourt and Soland (1971), for example, assume that area defenses can be exhausted through the sacrifice of a predictable number of attackers. It should not be forgotten that the defense may have better options than employing the principle.

It could be that neither attacker nor defender is able to predict the actions of the other, in which case a two-person zero-sum game results, possibly a Blotto game. Chapter 6 includes a discussion of defense problems viewed in this manner.

Defensive problems become especially complex when the targets, the interceptors or the attackers are of diverse types. Recent attempts to deal with this complexity are by Bertsekas et al. (2000), who apply an approximate kind of dynamic programming that has the virtue of suffering less from Bellman's curse of dimensionality, and by Brown et al. (2005), who solve a two-sided model using a large mixed-integer linear program. The latter is made the basis of a case study in Chapter 6.

Exercises

(1) The defender has 13 interceptors and wishes to conduct a defense that will maximize the cost of killing a target by attackers whose probability of killing the target is 1, if not intercepted. The interceptor kill probability is 0.5, and all interceptors for each attacker must be employed in a single salvo. What is the maximized cost, and how many interceptors should be assigned to the first, second, and third attackers? Use sheet "MaxCost" to find the answer. Ans. The maximized cost is 4.28 attackers. The first three attackers should be assigned 4, 3, and 3 defenders, respectively.

(2) The attacker kill probability is 0.7 and the defender kill probability is 0.6. The defense wishes to conduct a Prim–Read defense for which the maximum kill probability per attacker is 0.2. How many defenders are needed?
Ans. 5.

(3) Begin with the observation that $R' = \alpha R - \Delta$ and show by induction that formula (4.5) is correct.

(4) The maximum range of a ship's sonar for detecting incoming torpedoes is 1950 m. A battery of anti-torpedo-torpedoes (ATTs) is available to defend the ship against attacking torpedoes that approach at a speed of 10m/s. The ATT speed is 5 m/s. It takes 10 s to determine the outcome of any engagement, and any torpedo that gets within 2 m of the ship will detonate.

(a) How many salvos are possible without risking a detonation?
Ans. The sequence of launch ranges is 1950, 600, 150, 0. The launch at 150 will intercept at range 50, which is larger than 2, so three salvos are possible.

(b) Suppose that the ship has eight ATTs and that four torpedoes attack. Each torpedo will kill the ship with probability 0.5, if not intercepted. Each ATT will kill its torpedo target with probability 0.8, regardless of salvo index. If the ATTs are employed optimally, what is the ship's survival probability? Ans. 0.9945. Use sheet "DPDefender," and remember that a DP analysis for 12 ATTs and seven torpedoes will also solve all smaller problems in the process.

(5) The "DPDefender" sheet of *Chapter3&4.xls* implements (4.4) for up to 7 attackers, 12 defenders, and 3 stages, with the p and ρ parameters of Section 4.1 being inputs (there is a separate p for each stage). The optimized function $F_n(s,t)$ is shown in yellow cells, and the optimizing allocation of defenders to targets $x_n(s,t)$ is shown in green cells. Solve several problems and state whether you think each of the following propositions is true or false when all stages have the same p:

(a) For fixed (n,t), $x_n(s,t)$ is always a nondecreasing function of s. "The more you have, the more you use."

(b) All attackers are treated equally; that is, $x_n(s,t)$ is always an integer multiple of t.

(c) If $s \geq t$, then $x_n(s,t) \geq t$. "If resources permit, shoot at each target at least once."

(d) For fixed (n, s), $x_n(s,t)$ is a nondecreasing function of t. "The more targets there are, the more interceptors you use."

Also, give a qualitative description of what happens if the last stage has a higher p than the other 2, as might happen if defenders were more effective at close range.

(6) Suppose that the number of attackers A is random, with $P(A = i) = 0.1, 0.3, 0.4, 0.2$ for $i = 0,1,2,3$. The attackers, whatever their number, arrive sequentially, so that interceptors must be allocated to each attacker as it appears. Each interceptor kills its target with probability 0.5. Using (4.6), determine the firing schedule that maximizes the probability of destroying all attackers if there are four interceptors in total. One way to solve this problem is to construct a spreadsheet that solves the problem in general for at most three attackers.
Ans. Set p=1, since this makes the probability of survival equal the probability of killing all attackers. Then $m_1 = 2$, $m_2 = 2$, $m_3 = 0$. The probability of destroying all attackers is $F(4,0) = 0.55$. Using all four interceptors on the first two attackers leaves the defender helpless against a possible third attacker, but doing so is nonetheless optimal. Intuitively, the chance of having three attackers is so low that it is not wise to protect against the possibility. One possible spreadsheet solution is sheet "Exercise 6" of *Chapter3&4.xls*.

(7) Section 4.2.3 includes an example where $\rho = 0.5$, $p = 0.8$, and $m = 11$. Would the defender prefer 4 perfect ($\rho = 1$) interceptors to 11 imperfect ($\rho = 0.5$) ones

a) Using the measure of effectiveness $c(m)$?

b) Using the measure of effectiveness ρm, the "average number of kills?"

Ans. The defense prefers perfect interceptors in (a), but not in (b). Note the high emphasis that $c(m)$ places on interceptor quality.

(8) Using a modified version of (4.7), compute $c(8)$ under the assumptions that $p = 0.8$, $\rho = 0.5$, and that at most two interceptors can be allocated to any attacker. Sheet "MaxCost" cannot be used to answer this question, since it lacks a constraint on salvo size.
Ans. $c(8) = 3.49$.

(9) Continue the sample analysis begun in Section 4.2.4 by next guessing $\lambda = 0.15$. You should find that the required firing schedule utilizes 11 interceptors. Compare your calculations with the firing schedule obtained in

Section 4.2.3 for 11 interceptors. The resources are identical in the two analyses, but the optimization criterion differs.

(10) Suppose you must locate anti-aircraft defenses around a depot that may be attacked by enemy aircraft. The Near defense will always shoot down the attacking aircraft, but only after the aircraft has already dropped its bombs. In the Far defense, the probability of killing an aircraft is only 2/3, but the advantage is that any aircraft killed will not be able to drop their bombs first. You decide to arrange the defenses to maximize the average cost, in terms of enemy aircraft, of dropping a load of bombs.
(a) What do you think of the criterion?
(b) Which is better, the Near defense or the Far defense?
Ans. The cost of a successful attack is 1 with the Near defense or 2 with the Far defense, so the Far defense is better. See Example 4.

Chapter 5
Attrition Models

Another such victory over the Romans, and we are undone.

Pyrrhus

5.1 Introduction

Much of combat is about the destruction of enemy forces. The models in this chapter ignore all other aspects, including important aspects such as morale and maneuver. Victory is a matter of outlasting the enemy.

Most attrition models are descriptive in the sense that they are not built to optimize any particular tactical decision. The model simply describes how the numbers of the various forces involved will fluctuate with time, generally decreasing until some battle termination criterion is met. The purpose of the model may be to see how the evolution of the battle depends on the initial numbers and lethality of the various weapon systems involved, or to plan logistics or medical support.

At their simplest, aggregated attrition models can be easily implemented in spreadsheets, as will be shown by example in *Chapter5.xls*. More detailed high-resolution models may be tailored to specific terrain, recognize whether there is actually a line of sight from shooter to target, incorporate weather and time-of-day effects, and take hours to run on powerful computers. In this chapter we deal mainly with aggregated models that ignore most of these details.

An early application of attrition models was to air combat between fighters in World War One, and Hughes (2000) states that attrition models are particularly applicable to naval combat. Nonetheless, the main applications have historically been to land combat.

5.2 Deterministic Lanchester Models

Definition: A deterministic Lanchester model is a mathematical model composed of a system of ordinary differential equations (ODE). The ODE involve state variables that represent the number of entities of various types present in a battle, and each of the ODE expresses the time rate of change of one of these variables as a function of the others.

Lanchester (1916) applied ODE to situations where mutual attrition occurs continuously in time. He was inspired by air combat between fighters in World War One, and particularly wanted to investigate the importance of concentration

A. Washburn, M. Kress, *Combat Modeling*, International Series in Operations Research & Management Science 134, DOI 10.1007/978-1-4419-0790-5_5,

of forces. This class of ODE-based combat models is named after him, even though Ossipov (Helmbold and Rehm, 1995) is equally deserving.

The general idea in a Lanchester model is to define some state variables representing force levels, determine the ODE for the rate of change of each of these variables, and then solve the resulting equations as a function of time. In the case of air combat, let the state variables $x(t)$ and $y(t)$ be the number of surviving blue and red aircraft at time t on each of two sides, and then hypothesize that the rate of change of each variable is proportional to the other. Symbolically, we have

$$\frac{dx(t)}{dt} = -ay(t);\ x(t) > 0 \text{ and } \frac{dy(t)}{dt} = -bx(t);\ y(t) > 0\,. \tag{5.1}$$

The proportionality constant a represents the number of blue aircraft shot down by each surviving red aircraft per unit of time, and similarly for b with the roles reversed. Each equation holds only when its state variable (the one whose time derivative is shown on the left-hand side) is positive; once the variable reaches 0, it ceases to decrease because the other side has nothing left to shoot at.

Given initial values for all state variables at time 0, a set of ODE determines the state variables for all future time. Sometimes (as in the case of (5.1)) the equations can be solved analytically, but in general a numerical solution will be necessary. Although there are more efficient methods, the only numerical method to be used here is Euler's method where time is made discrete and the differential equations are replaced by difference equations. If the time increment is Δ, the difference equation counterpart to (5.1) is

$$x(t+\Delta) = x(t) - a\Delta y(t) \text{ and } y(t+\Delta) = y(t) - b\Delta x(t). \tag{5.2}$$

Since the initial values $x(0)$ and $y(0)$ are given, (5.2) determines $x(\Delta)$ and $y(\Delta)$ by substituting into the right-hand side. With $x(\Delta)$ and $y(\Delta)$ known, (5.2) determines $x(2\Delta)$ and $y(2\Delta)$, and so on until a termination condition is reached. The termination condition might be that a certain amount of time has expired or that the force ratio has increased or decreased to the point where one side or the other would surrender. If any computation produces a negative value for a state variable, then the state variable is instead taken to be 0. The theory of ODE guarantees that (at least for all of the ODE considered here), in the limit as Δ approaches 0, the difference equation solution approaches the true solution of the ODE.

The reader may wish to experiment with sheet "SquareLaw" of *Chapter5.xls*, which displays both numerical and analytic solutions of (5.1) (the analytic solution is given in Exercise 14). The analytic solution is actually the solution of (5.1) without the positivity restriction, so it must be used with care. When (say) red becomes negative, the analytic formula for the number of surviving blues will actually increase with time, as if red were somehow "unshooting" at blue. The numerical solution avoids this unwelcome possibility by replacing any tentative negative number of survivors with 0.

Although the time increment Δ in (5.2) is meant to be an arbitrary small number, sometimes a more thoughtful choice may lead to a more valid representation of reality. This is particularly true if combat is conducted in a sequence of powerful discrete attacks, as at theWorld War Two battle of Midway where strikes were conducted by waves of carrier-based attackers. Hughes, for example, has employed his "salvo equations" extensively in evaluating naval combat (see his chapter in Bracken et al. (1995), or Lucas and McGunnigle (2003)). In that case Δ represents the length of a combat cycle, and (5.2) represents the idea that the strikes in each cycle are simultaneous – the salvos cross in midair.

If red strikes blue first in each cycle of combat, it may be best to base the update for $y(t+\Delta)$ in (5.2) on $x(t+\Delta)$, rather than $x(t)$. This would make little difference in solving ODE when Δ is small, but could be a substantial advantage to red if $x(t+\Delta)$ is significantly smaller than $x(t)$. Similarly, if blue strikes red first, the update for $x(t+\Delta)$ might be based on $y(t+\Delta)$, rather than $y(t)$. Exercise 4 shows that the order of making updates can be significant.

Since all state variables are functions of time (t), and since all derivatives are with respect to time, it is customary to write ODE like (5.1) in abbreviated form where the time argument is omitted on the right-hand side, as we will do in the rest of this chapter.

When ODE involve only two state variables, it may be useful to eliminate time by taking the ratio of derivatives. The result in the case of (5.1) is

$$\frac{dy}{dx} = \frac{bx}{ay};\ x > 0 \text{ and } y > 0. \tag{5.3}$$

By separating variables, we discover that the solution of (5.3) is $y^2 = (b/a)x^2 + (\text{constant})$, as can be verified by differentiation. Since this must also hold at time 0, when x and y are known to be (say) x_0 and y_0, respectively, the constant of integration must be such that

$$a(y^2 - y_0^2) = b(x^2 - x_0^2)\,. \tag{5.4}$$

Equation (5.4) is a relation between x and y that holds throughout the period where both state variables are nonnegative and can be used to predict which side will win a battle that ends when one side is reduced to 0. If the "fighting strengths" of the two sides are defined to be bx_0^2 and ay_0^2, respectively, then whoever has the larger fighting strength will win such a battle, and (5.4) can be used to predict the numbers of survivors. Inserting 0 for x in (5.4) and solving for y, for example, we find that $ay^2 = ay_0^2 - bx_0^2$, which produces a positive number of red survivors (y) as long as red has the larger fighting strength. The dependence of the fighting strength on the square of the number of units is responsible for the term "square law" in referring to (5.1).

Example 1: Suppose $x_0 = 100$, $y_0 = 200$, $a = 0.03$, and $b = 0.01$. The fighting strengths of blue and red are 300 and 400, respectively, so red will win. Figure 1 shows the course of the battle by plotting y versus x as determined by (5.4). Not only does red win, but he has half of his initial force left when blue is eliminated. See sheet "RedBlue" of *Chapter5.xls* for variations.

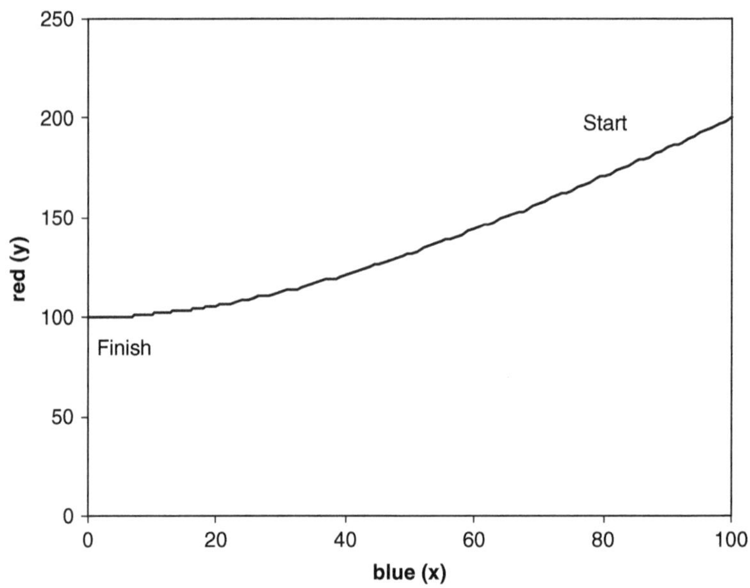

Figure 1: The course of a square law battle with time eliminated.

In Example 1, blue loses in spite of the fact that he is only outnumbered by a factor of 2, while each of his units is three times as lethal as one of red's. The reason for this is that fighting strength is proportional to the square of the initial number, but only to the first power of the lethality coefficient. A small force of excellent units may very well lose to a larger force of conventional units. This emphasis on numbers has a simple intuitive explanation. Intuitively, there are two reasons for blue unit Tom to desire an additional friendly unit Dick in a square law battle. One reason is that Dick will shoot at the enemy Harry, perhaps killing Harry before Harry can kill Tom. The other reason is that Harry may shoot at Dick instead of Tom; that is, the additional unit will dilute the enemy's fire. An increase in the lethality coefficient fulfills the first function, but not the second, so (given a choice) it is better to increase the initial number of units. If the square law does indeed represent combat, this is an important observation about the kind of forces that ought to be fielded.

The square law is sometimes referred to as "aimed fire" because the rate at which blue (say) loses units has nothing to do with x, the number of blue units. It is as if red units never had to search for blue targets to shoot at. If blue targets were hard to find or if red were simply firing into an area without aiming at any specific blue unit, we would expect blue losses to be proportional to x, as well as y. The resultant "unaimed fire" ODE are

$$\frac{dx(t)}{dt} = -axy \text{ and } \frac{dy(t)}{dt} = -bxy. \tag{5.5}$$

It is not necessary to restrict (5.5) to nonnegative values of x and y – negative values are avoided automatically because each time derivative becomes 0 when its state variable becomes 0. As in the case of the square law, there are corresponding difference equations that can be used to solve (5.5) numerically (see Exercise 2).

The parameters a and b do not have the same meaning as in the square law, so it does not make sense to compare solutions of (5.4) and (5.5) with the intention of determining whether aimed or unaimed fire is "better" for one side or the other.

Elimination of time in (5.5) by taking the ratio of derivatives leads to the equation $dy/dx=b/a$, which has the solution

$$a(y - y_0) = b(x - x_0)\,. \tag{5.6}$$

The fighting strength is now bx_0 for blue or ay_0 for red, the product of initial number of combatants and lethality in each case. The unaimed model is therefore sometimes called the "linear law," since only the first power of initial number is involved. Initial numbers and lethality are emphasized equally in the linear law, so it is easier to make up for being outnumbered by increasing unit lethality.

The literature contains many generalizations and different kinds of "laws" besides square and linear. Deitchman (1962), for example, considers a model of guerilla warfare where one side (the guerillas) uses aimed fire while the other uses area fire. Cases intermediate between area fire and aimed fire have been considered, and there is no reason why both types of fire should not be present at the same time, as they will be below in Section 5.6. There may be multiple entity types (infantry and artillery, for example) on each side. Our next example is of this type.

Example 2 (Battle of the Atlantic): This is a roughly realistic model of the World War Two U-boat war in the Atlantic. We assume that German submarines operate only in "wolf packs" and that all Allied ships sail in escorted convoys. Morse and Kimball (1950) make the approximation that $5n/c$ and $nc/100$ merchant ships and U-boats, respectively, will be lost when a convoy defended by c escorts is attacked by n submarines. These approximations were arrived at statistically, after examining the results of previous convoy encounters. We retain that assumption. Since the exchange ratio of U-boats lost to escorts lost in convoy engagements was about 5/1, we will also assume that the average number of escorts lost is $nc/500$. Other numbers assumed below were typical of the time. Let

M	$\equiv$	cumulative merchant ships sunk (state variable)
S	$\equiv$	remaining submarines (state variable)
E	$\equiv$	remaining escorts (state variable)
p_c	$\equiv$	probability that a convoy is attacked by a given wolf pack = 0.01
t_c	$\equiv$	time a convoy must be escorted (both directions) = 30 days
T_c	$\equiv$	cycle time for an escort from one convoy to the next = 50 days
f_c	$\equiv$	fraction of escorts escorting convoys = $t_c/T_c = 0.6$
t_s	$\equiv$	patrol time for a submarine = 20 days
T_s	$\equiv$	cycle time for a submarine from one patrol to the next = 50 days
f_s	$\equiv$	fraction of submarines on patrol = $t_s/T_s = 0.4$
r	$\equiv$	rate at which convoys leave = 1/day
m	$\equiv$	convoy size = 40
n	$\equiv$	wolf pack size = 4
p_s	$\equiv$	probability of sub loss per cycle due to other causes other than convoy engagements = 0.04
R_s	$\equiv$	rate of replacement of submarines (new construction) = 0.7/day
R_E	$\equiv$	rate of replacement of escorts (new construction) = 0.5/day

We assume that the following three formulas hold

Sf_s/n=		number of wolfpacks on patrol
e	$\equiv$	engagement rate counting engagements both ways = (convoy sailing rate)(p_c)(wolfpacks on patrol) = $(2r)(p_c)(Sf_s/n)$
c	$\equiv$	escorts/convoy = $(Ef_c)/(rt_c) = E/(rT_c)$

There was a debate within the German navy about whether convoys returning from Britain to the United States should be engaged, the argument for not doing so being that the returning ships were empty. Here we are assuming that engagements happen in both directions, hence the factor of 2 in the formula for e. Given all the above assumptions, the ODE for the three state variables are

$$\frac{dM}{dt} = e(5n/c) = 10r^2T_c p_c f_s (S/E);\ M > 0,$$

$$\frac{dS}{dt} = -e(nc/100) - p_s S/T_s + R_s = -S([2p_c f_s/(100T_c)]E + p_s/T_s) + R_s;\ S > 0,$$

$$\frac{dE}{dt} = -e(nc/500) + R_E = -S[2p_c f_s/(500T_c)]E + R_E;\ E > 0.$$

These ODE exhibit several new features. M starts at 0 and can only increase, since the cumulative losses of merchant ships can only go up. M plays no role in the ODE for S and E and serves only as a counter for the progress of the battle. Note that the rate of loss of merchant ships is proportional to the square of the sailing rate and has nothing to do with the size of the convoy. If one were to double the convoy size and halve the sailing rate, merchant losses would go down sharply because there would be fewer engagements and also fewer losses per engagement. This was basically the observation that led to increased convoy sizes in World War Two. The ODE for S and E involve both positive and negative terms, since submarines and escorts were both sunk and replaced.

The wolf pack size does not appear in any equation, which is superficially at odds with histories of the battle that report devastating effects of wolf pack introduction. However, while the effect of encountering a wolfpack was indeed devastating for the unlucky convoy, the effect is exactly balanced in these equations by the encounter rate's inverse proportionality to wolf pack size. McCue (1990) includes a more in-depth analysis of this issue.

Substituting numbers, with time in days,

$$\frac{dM}{dt} = 2S / E;\ M > 0,$$

$$\frac{dS}{dt} = -S(1.6\times10^{-6} E + 8\times10^{-4}) + 0.7;\ S > 0,$$

$$\frac{dE}{dt} = -3.2\times10^{-7} ES + 0.5;\ E > 0.$$

Figure 2 shows the solution of these equations for the initial conditions $M(0) = 0$, $S(0) = 50$, $E(0) = 100$, except that merchant losses are divided by 10 to keep all curves on the same scale. During the 2000 day period, there are $10 \times 250 = 2500$ merchant ships sunk. The number of escorts grows at approximately 0.5/day throughout the period, since escort attrition is negligible. The number of submarines eventually reaches a peak and declines, but not as catastrophically or as fast as actually happened. A major reason for this is that p_s actually increased strongly with time in WWII, whereas it has been held constant here.

The "Atlantic" sheet of *Chapter5.xls* (from which Figure 2 is taken) permits the user to experiment with different initial conditions or parameters. A great many excursions are possible; e.g., change p_c to indicate better submarine sensors or change R_E to illustrate the benefit of increased construction capacity.

The actual Battle of the Atlantic exhibited periods of success for the two sides as various secret codes were broken and technological innovations were introduced. It also had strong geographical influences. Our simple Lanchester model misses all of this interesting detail, but it does make one thing clear: as long as convoy engagements are well approximated by the Morse and Kimball equations for losses in convoy engagements, the submarines are doomed in the long run because they cannot sink escorts fast enough.

Lanchester models can have hundreds of state variables, as in Section 5.6 below. They can also have embedded optimization problems – Should blue's artillery shoot at red's infantry or at red's artillery? Taylor (1983) is an excellent introduction to the rich selection of possibilities that the ODE paradigm makes possible. Since our intention here is not to be exhaustive, we pass on to the consideration of the stochastic version.

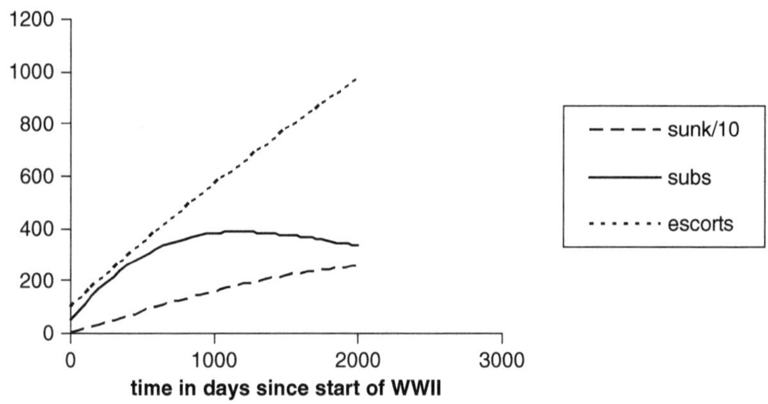

Figure 2: The Battle of the Atlantic. Merchant ships sunk are divided by 10.

5.3 Stochastic Lanchester Models

In solving Lanchester's ODE using Euler's method (Equation 5.2), the time increment Δ is ideally a small number. If Δ is small, the number of casualties in one time increment may be similarly small, say 0.1. What does it mean to have 0.1 casualties? It seems illogical to model something that has to be an integer with a real number that could very well be fractional, as is the case when ODE are employed. There may be no harm in the practice if the forces engaged are sufficiently large, but how are we to know what size is "sufficiently large" unless we are able to model the battle with integers in the first place? In fact, why is it that something as laden with chance events as a battle would be modeled with any kind of a deterministic model? All of these questions lead to the employment of stochastic models.

One of the charms of modeling battles with Lanchester's ODE is that they permit a stochastic interpretation that is responsive to criticisms such as those expressed above. Letting $\mathbf{x} = (x_1,\ldots,x_k)$ stand for a vector of k state variables, the general form of ODE is

$$\frac{dx_i}{dt} = -f_i(\mathbf{x});\, i = 1,\ldots,k\,. \tag{5.7}$$

Most of the examples considered so far have had $k = 2$, but that restriction is not essential. The deterministic interpretation is that $f_i(\mathbf{x})$ is the rate at which x_i decreases. In contrast, the stochastic interpretation is that x_i decreases in a small time Δ by a single unit, if at all, with the probability of that event being $\Delta f_i(\mathbf{x})$. To state the stochastic interpretation more precisely, $f_i(\mathbf{x})$ is taken to be the transition rate of a continuous time Markov chain (Ross, 2000).

The same functions underlie either interpretation, so the stochastic interpretation requires no new data. Since the same data underlie either interpretation, we use the term "Lanchester model" to refer to either one. In the stochastic version, the *average* losses to x_i in time Δ are $1(\Delta f_i(\mathbf{x}))+0(1-\Delta f_i(\mathbf{x}))=\Delta f_i(\mathbf{x})$, which agrees with the deterministic version. We can therefore see that the deterministic version is obtained from the stochastic version by employing EVA, as described in Chapter 1. In fact, comparing the two versions is a good opportunity to study the effects of EVA.

Example 3: Consider a square law battle (Section 5.2) where $x_0 = 10$, $y_0 = 8$, $a = 0.9$, and $b = 0.8$. The deterministic theory predicts that blue will always win with about 5.29 survivors. Under the stochastic interpretation, red wins about one fourth of the time, generally when he gets lucky early. This can be demonstrated in a Monte Carlo simulation (Appendix C) where time is repetitively advanced by Δ until one side or the other is wiped out, with survivors being decremented in each step based on whatever random numbers are generated. See sheet "SquareLawRnd" of *Chapter5.xls* for a demonstration. If you use *SimSheet.xls* (Appendix C) to perform lots of replications, you will see that the distribution of the net number of survivors $(x - y)$ is bimodal, with a positive mode where blue wins and a negative mode where red wins. Battles where both sides are nearly wiped out are rare. The average net number of blue survivors is about 3.4.

Example 3 shows that the deterministic and stochastic versions of a small Lanchester model can differ strongly. Evidently, the EVA shortcut can cause significant errors in small battles.

Although Monte Carlo simulation is one way to explore the stochastic interpretation, analytic computations based on the Chapman–Kolmogorov equations (Morse and Kimball, 1950; Ross, 2000) are also possible. Consider any battle where each of two sides is represented by one state variable ($k = 2$). To emphasize the integer nature of all state variables, refer to the state as (m,n) rather than (x_1, x_2). If the state is currently (m, n), then the predecessor state must have been either $(m, n + 1)$ or $(m + 1, n)$, and the successor state must be either $(m - 1, n)$ or $(m, n - 1)$. Figure 3 is part of a state transition diagram showing both of the ways in which the state might get to (m, n), as well as both of the ways in which the state might leave (m, n). The transition rates between these states are known. For example, the rate of going from $(m, n+1)$ to (m, n) is $f_2(m,n+1)$, the rate at which the second state variable decreases from state $(m, n+1)$. The Chapman –Kolmogorov equations are ODE involving these rates, with the probabilities of being in the various states at time t as state variables. Specifically, the equation for $P(m,n,t)$, the probability of being in state (m, n) at time t, is

$$\frac{dP(m,n,t)}{dt} = f_1(m+1,n)P(m+1,n) + f_2(m,n+1)P(m,n+1) \\ -(f_1(m,n) + f_2(m,n))P(m,n,t). \tag{5.8}$$

Equation (5.8) states that the probability of being in state (m, n) at time t increases because of transitions into that state (the first two terms on the right-hand side), and decreases because of transitions out (the last term). There is a similar equation for every possible state. State (0,0) will never be observed because the battle stops as soon as one side or the other is reduced to 0. Therefore, if the battle starts with m_0 blue units and n_0 red units, there are a total of $m_0 n_0 + m_0 + n_0$ equations. Some of these equations omit some of the terms in (5.8) – the first term is missing from the equation for state $(m_0, 0)$, for example, since state $(m_0 + 1, 0)$ is impossible.

At time 0, all of the probabilities $P(m,n,0)$ are 0 except for $P(m_0, n_0, 0)$, which is 1.0. Since all probabilities are known at time 0, Euler's method can be used to find all probabilities at time Δ, and then at time 2Δ, etc. At the end of the implied long computational process, we will know the probability of being in every state at every time. Contrast this with the simpler deterministic interpretation, where we discover how the state of the system evolves by solving a system of only 2 ODE. Obviously the stochastic interpretation leads to the need for considerably more computation than the deterministic interpretation. Nonetheless, in spite of the computational difficulty, the Chapman–Kolmogorov equations are solvable in problems that are not too large.

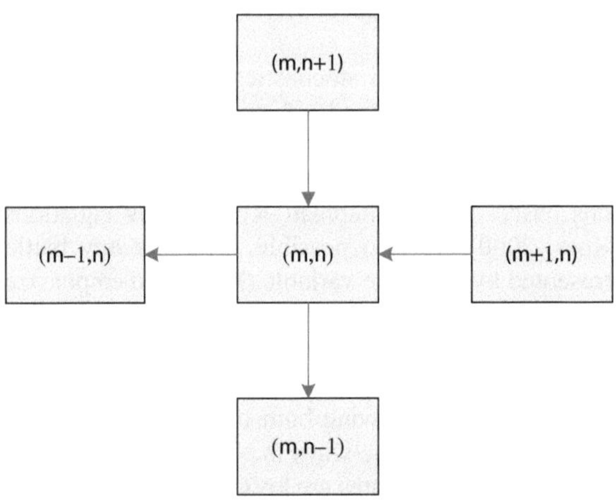

Figure 3: Transitions into and out of state (*m,n*).

Example 4: Consider a square law battle somewhat larger than the one considered in example 3, with $m_0 = 20$, $n_0 = 40$, $a = 0.01$, and $b = 0.02$. Thus $f_1(m,n) = 0.01n$ and $f_2(m,n) = 0.02m$. The deterministic version has red winning at time 62.32 with 28.28 survivors. Figures 4 through 6 display the actual probabilities of being in the various states at times 20, 40, and 60. The state is somewhere around (14, 35) at time 20, with little chance of the battle being over by then. A t time 40, some of the probability has migrated to the m=0 axis, where red has already won the battle. By time 60, most of the probability is on that axis. By time 80, nearly all of the probability would be on that axis, with red having about 30 survivors. The sequence of pictures creates the impression of a probability mass that starts at (20,40) and spreads out as it makes its way over to the axis where $m = 0$, to which it sticks. At all times the total mass (volume under the bulge) is 1.

Once the probability distribution $P(m,n,t)$ is known, various derived quantities such as the expected number of red survivors at time t can be easily computed.

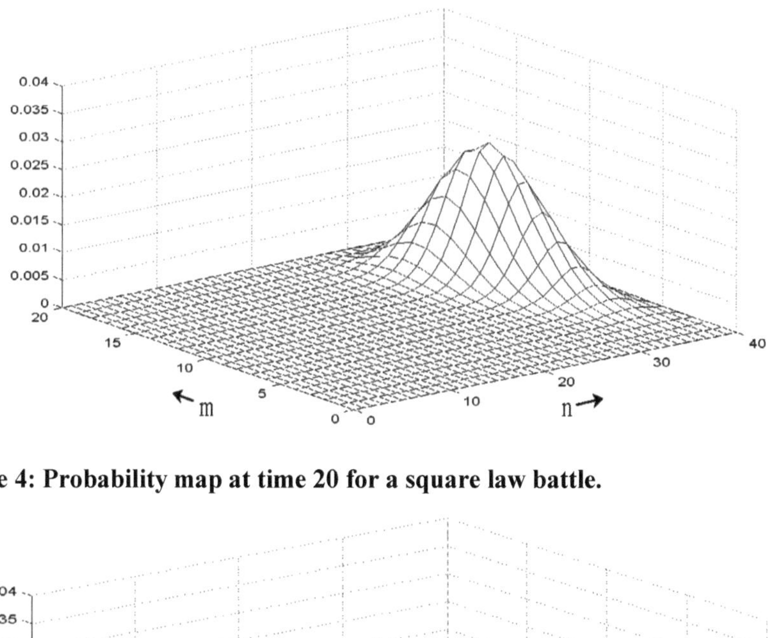

Figure 4: Probability map at time 20 for a square law battle.

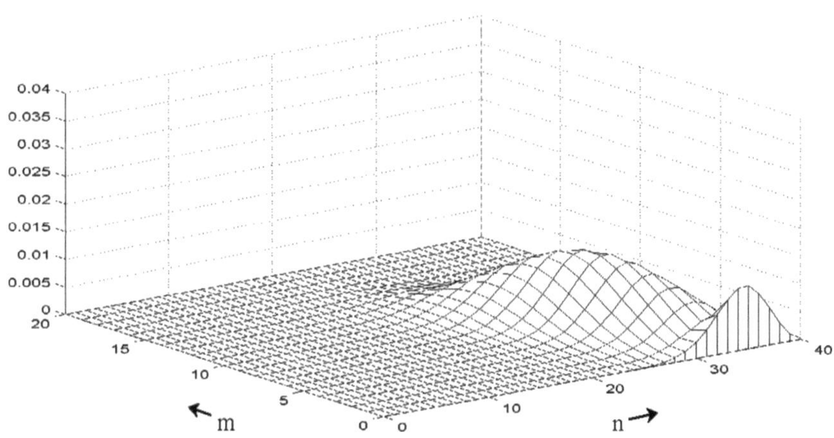

Figure 5: Probability map at time 40 for a square law battle.

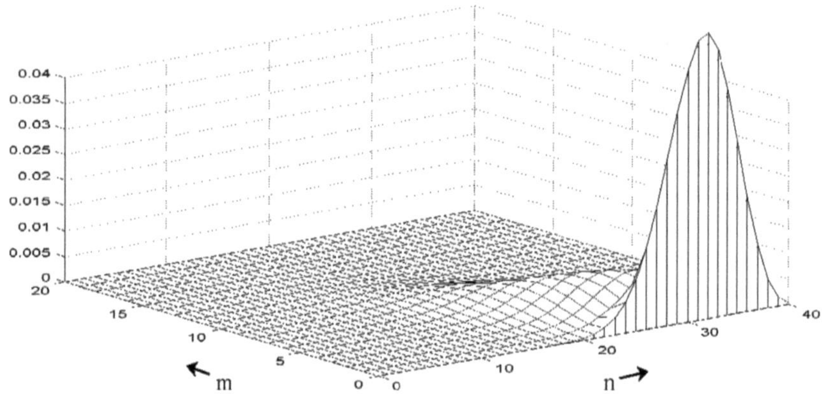

Figure 6: Probability map at time 60 for a square law battle.

As in the deterministic case, it may be useful to eliminate time in models where there are only two state variables. The crucial observation is that state (m, n) will eventually be succeeded by either state (m, n–1) or state (m–1, n). According to Markov chain theory (Appendix A), the probability of the latter is

$$Q(m,n) = \frac{f_1(m,n)}{f_1(m,n) + f_2(m,n)}, \tag{5.9}$$

with $(1 - Q(m,n))$ being the probability of the former. These probabilities can then be the basis of a Markov chain where time is not involved. They are particularly useful if the object of analysis is to evaluate a measure of effectiveness (MOE) that depends only on the terminal state of the battle. Such measures include "the number of blue survivors" and "the probability that blue wins a fight to the finish". Let $MOE(m,n)$ be the expected value of this terminal measure, given that the state is currently (m, n). Then the following equation connects $MOE(m,n)$ to corresponding values at each of its successor states:

$$MOE(m,n) = Q(m,n)MOE(m-1,n) + (1 - Q(m,n))MOE(m,n-1). \tag{5.10}$$

Now, $MOE(m,n)$ is assumed to be known in terminal states. For example, if $MOE(m,n)$ is the probability that blue wins a battle to the finish, then $MOE(m,0) = 1$ and $MOE(0,n) = 0$. Since $MOE(m,n)$ is known at the terminal states, equation (5.10) can first be used to evaluate the function in states that lead only to terminal states. After sufficiently many iterations of (5.10), $MOE(m,n)$ can be evaluated in any state. The computations are efficient, compared to Monte Carlo simulation or to first employing the Chapman–Kolmogorov equations to find the state probabilities at some time large enough to guarantee battle completion (perhaps time 100 in Example 3). They can be easily organized in a spreadsheet.

Example 5: Consider again Example 3, but now our object is compute the exact probability that blue wins, and also the average net excess of blue survivors. To find these quantities, we employ sheet "Endgame" of *Chapter5.xls*, which is set up to do the calculations for square law battles that proceed until one side is annihilated. We first insert 0.9 and 0.8 for a and b (or alpha and beta, as they are called on that sheet). We next fill out the green border (terminal) cells of the computational rectangle with the appropriate values. To find the probability that blue wins, insert 1 in the row where blue wins, and 0 in the column where red wins. Make no changes to the interior of the rectangle, which is composed entirely of formula (5.10). In those interior cells, $Q(m,n)$ is called *leftprob*(m, n) because the state moves left when blue loses a unit (VBA code in Module 1 implements *leftprob*(m, n) as in (5.9)). The probability that blue wins from state (10, 8) turns out to be 0.78. To instead find the net excess of blue survivors, insert m in the row where blue wins and $-n$ in the column where red wins. The net excess of blue survivors from state (10, 8) is 3.38. These exact numbers are comparable to those found earlier in Example 3 by Monte Carlo simulation.

The deterministic solution is not a good approximation of the stochastic solution in Examples 3 and 5, since winning the battle is very much subject to chance. The counterparts to Figures 4, 5, and 6 would show an accumulation of probability on both axes, rather than only one. The deterministic version does better in Example 4, as we might expect because the numbers of units engaged are larger, but even in Example 4 the number of red survivors is significantly random, rather than always 28.28 as predicted deterministically (it is never 28.28, in fact, since it is always an integer).

In what sense, then, is the deterministic model an approximation of the stochastic model? Based on the examples considered so far, we might offer the following answer: In situations where large numbers of units are involved on both sides, and where one of two sides is clearly stronger than the other, the number of survivors in the deterministic model is nearly the expected value of the number of survivors in the stochastic model. This is not an entirely satisfactory answer, since words like "large" and "nearly" are not well defined, but at least it explains why the deterministic model typically forecasts a number of survivors that is not an integer. It also has the advantage of being in accord with experience. Taylor (1983, Section 4.16) summarizes that experience and gives several additional references.

The statement that the deterministic version of a Lanchester model produces the expected value of the stochastic version is only approximately correct. The approximation is better in some circumstances than others, but at least the error is quantifiable. There are other deterministic models, not of the Lanchester type, that produce fractional answers to problems involving discrete participants, but where there is no associated stochastic interpretation that can serve as a test of ultimate truth. Such models should be used with caution, if at all.

We have discussed two different interpretations of a Lanchester model. There is a third interpretation that should be avoided. This is the one where Euler's method is used to solve the deterministic version, but the number of survivors is rounded to an integer after every time step. This achieves a cosmetic goal at the cost of destroying the integrity of the deterministic solution. The object ought to be to solve the ODE accurately, rather than to enforce an integer property that (in the deterministic version) is not appropriate.

5.4 Data for Lanchester models

Lanchester models require that the analyst know attrition coefficients of various kinds. There may be only two such parameters in simple cases such as the square and linear laws, but the number of such parameters grows fast with the number of state variables. There are basically two methods that have been used to find such data, the engineering approach and the statistical approach. The two methods are not mutually exclusive; the Battle of the Atlantic (Example 2) employs both of them. In this section we take a closer look at the methods.

5.4.1 *The Engineering Approach*

In this approach, the unknown parameter is expressed as a function of other known or more easily measurable parameters, with the function depending on assumptions made about the battle.

Consider Lanchester's linear law, the law of "unaimed fire." The parameter a is the rate at which blue casualties occur per blue unit, per red unit. To estimate a, we might reason that red is firing into an area of size A without knowing exactly where any blue unit is located. If the lethal area (formula 2.3) of each red round is L, and if there are x blue units located within A, then the average number of casualties per randomly fired round will be $x(L/A)$. If each red unit fires rounds at rate r, and if there are y red units, then the total rate of blue loss is $rxy(L/A)$. The attrition coefficient a is therefore rL/A, the coefficient of xy, in units of inverse time. Parameter a has been expressed in terms of three more easily measured quantities.

Alternatively, we might assume that red *does* aim at individual blue units, but that each red unit must search the area A randomly to find his next target. If each red unit searches area at rate S, then red will find blue targets at rate xS/A. The total red force will therefore find blue units at the combined rate xyS/A. If the time to destroy a blue unit is negligible, we then have Lanchester's linear law with a being S/A, again in units of inverse time (this illustrates that the abstract name "linear law" should be preferred to the more evocative "unaimed fire," since here we have an example where the linear law applies where fire is aimed). If the destruction time is not negligible, call it τ. Since the average time to find a blue unit is the reciprocal of the rate of finding blue units, the total time for a red unit to dispose of a blue unit, counting both search and destruction, is $\tau + A/(xS)$. The total attrition rate is then $y/(\tau + A/(xS))$, or $xyS/(A + x\tau S)$. This expression acts like the linear law when x is large (attrition is limited by the time for destruction) or the square law when x is small (attrition is limited by the time for search). This is neither the square law nor the linear law, but rather a new, more general law. Whatever the law is called, the attrition rate can still be quantified by measuring more basic parameters. Brackney (1959) explores the behavior of this model in detail.

The engineering approach is particularly apt if the purpose of studying the Lanchester model in the first place is to determine the importance of certain parameters. Should current tanks be upgraded to include an advanced target acquisition system that will reduce the time τ needed to destroy a target? We might pose the specific question, "Would it be better to retain the current tank inventory with a better (smaller) value of τ or to spend the same amount of money buying more tanks of the current design?" This is a variation of the quality versus quantity question that Lanchester explored in World War I.

Example 6: Let τ be the mean time that it takes for a tank to destroy its target, and distinguish between hitting the target and killing it. Each hit kills the tank's target with probability P_K. Firing continues until the target is finally killed, but there may be multiple hits before that happens (we assume that if a hit on the target does not kill it, there is no damage). The time for the first shot and the probability of a hit on the first shot are τ_1 and P_1, respectively. The time for a shot following a miss and the probability of a hit on such a shot are τ_m and P_m, and the time for a shot following a hit and the probability of a hit on such a shot are τ_h and P_h. This is a complicated situation with seven inputs – the sequence of hits and misses might be a long one before the last hit finally kills the target. Nonetheless, the mean time to complete the process is known. Using Barfoot (1969), the mean time required to destroy a target is given by the formula

$$\tau = \tau_1 + \frac{1-P_1}{P_m}\tau_m + \frac{1-P_K}{P_K}(\tau_h + \frac{1-P_h}{P_m}\tau_m). \tag{5.11}$$

We forego the complete derivation, but (5.11) can be regarded as the sum of three terms. The first is the time until the first shot arrives at the target. The second is the additional time until the target is first hit (possibly zero – note that the term vanishes if P_1=1). The third term is the additional time between the first hit and the last (killing) hit, again possibly 0. Assuming that the time to find the next target is negligible, or a constant already included in τ_1, the corresponding attrition rate coefficient is simply the reciprocal of τ. Sheet "Tanks" of *Chapter5.xls* includes the computation of this formula in the context of a square law battle where you are able to influence the seven parameters that the blue versus red attrition rate depends on. Given a budget, you can decide whether to spend it on quality or quantity.

5.4.2 The Statistical Approach

The engineering approach has its virtues, but there are many possibilities for error. Reconsider what happens when one tank fires at another. Whether the next shot misses or not may depend on how big the previous miss distance was, rather than the simple fact of a miss, and may also depend on the distance between tanks and other variables not even considered in (5.11). Damage can accumulate, so that multiple hits result in a kill even though no individual hit would do so. A tank target can suffer a mobility kill while still being able to shoot, although it will of course be more vulnerable once it is immobile. Detailed, high-resolution simulations of tank battles acknowledge all of these phenomena, but the simple model lying behind (5.11) does not. Even if the model were fundamentally correct, there are still possibilities for error in estimating each of the seven parameters that are involved.

If data about tank battles is available, including attrition data, it is tempting to use it to estimate the attrition rate coefficients directly. Rather than try to explain how attrition happens, we can treat the process as a black box and deal with the phenomenon itself. This is the statistical approach. It applies to all kinds of attrition processes, not just to tank battles.

A frequently cited example of this approach is Engel's (1954) analysis of the battle of Iwo Jima, where an initial force of about 20,000 Japanese troops was eventually annihilated by the United States after about a month of combat. Only the starting level is known for the Japanese forces, since more detailed Japanese records did not survive the battle, but the number of US troops involved is known for every day of the battle. Engel's idea was to model the battle as one ruled by the square law, except for a correction term required by the multiple landings of US troops. There are only two unknown parameters, so why not search for the pair that does the best job of matching the known US force levels?

Figure 7 shows the result of Engel's matching. For the fit to US active troops, the two parameters were $a = 0.0544$ US casualties per Japanese man-day, and $b = 0.0106$ Japanese casualties per US man-day. The disparity between attrition rates might be explained by the fact that Japanese forces were fighting from well prepared defensive positions, but similar disparities have been found in battles where the roles are not so clearly defined (the Kursk battle considered below, for example). In spite of this disparity, casualties on the two sides were about equal. Essentially all 20,000 Japanese troops were killed. The US also had about 20,000 casualties, although only about a quarter of these died. Since there were about three times as many US troops as Japanese troops during the first week of the battle, and since the square of three exceeds the attrition rate ratio of 5, Lanchester's square law correctly predicts a US victory. The fit shown in Figure 7 is also remarkably good. Sheet "IwoJima" of *Chapter5.xls* replicates the Iwo Jima battle using Engel's estimated coefficients (see Exercise 7).

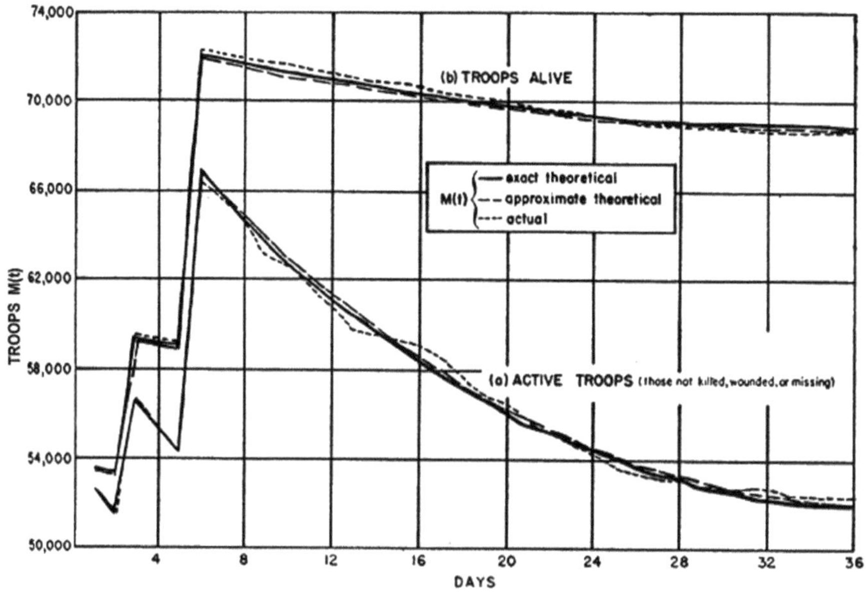

Figure 7: Engel's square law fit to the battle of Iwo Jima in World War II. The increasing portions correspond to US troop landings. There are actually two fits, depending on whether one counts US troops still alive or still active, with different attrition rates. In either case, it can be seen that Lanchester's square law does a good job of tracking the number of US troops.

Engel's analysis was performed well after the battle, but, in principle, the same attrition coefficients that applied to Iwo Jima might have been supposed to hold for other similar battles in World War II. In fact, analyses of other World War II battles reveal considerable consistency between coefficients, even between theaters. See Willard (1962) for analyses of battles going all the way back to the year 1618.

Most of the work that has been done fitting equations to battles is not so flattering to the Lanchester theory as that of Engel. An example of this is the battle of Kursk.

Example 7: The battle of Kursk was a tremendous 2-week battle in World War II that saw the German forces inside Russia switch from offensive to defensive tactics at about the eighth day. Both sides kept track of both the number of personnel available and the number of casualties on each day, data which have been collected in CAA (1998). Table 1 shows part of the data. Reinforcements are not shown, so troop levels are sometimes surprising – there are more German troops on day 9 than on day 8, for example. Nonetheless, if battles in the aggregate have the kind of predictability that Lanchester models attempt to exploit, we should be able to predict daily losses using a simple formula based on troop levels. The square law that Engel had such success which involves only two parameters, and there is enough data available to estimate them.

Time (days)	German level	German loss	Soviet level	Soviet loss
0	307,365	800	510,252	130
1	301,341	6,192	507,698	8,527
2	297,205	4,302	498,884	9,423
3	293,960	3,414	489,175	1,0431
4	306,659	2,942	481,947	9,547
5	303,879	2,953	470,762	11,836
6	302,014	2,040	460,808	10,770
7	300,050	2,475	453,126	7,754
8	298,710	2,612	433,813	19,422
9	299,369	2,051	423,351	10,522
10	297,395	2,140	415,254	8,723
11	296,237	1,322	419,374	4,076
12	296,426	1,350	416,666	2,940
13	296,350	949	415,461	1,217
14	295,750	1,054	413,298	3,260

Table 1: German and Soviet force levels and losses during the 14-day battle of Kursk . Day 0 starts at 1,800 on July 4, 1943. Losses listed for each day are during the 24 h preceding its start time. Force levels are as of the start time.

But fitting a simple formula will not be easy. Note the high Soviet losses on day 8, for example. Considering only the force levels involved, which differ little from the force levels on day 7 or day 9, it would be hard to explain how such a thing could occur. The historical explanation is that day 8 began with the initiation of a climatic German armored attack meant to close some pincers on the Kursk salient. The attack ultimately failed, but only after the determined Soviets had sacrificed the requisite forces to stop it. There were also tremendous losses of armor on both sides on that day. Part of the statistical problem is that massive attacks like the one on day 8 compress time in the sense that there is much more combat activity on some days than on others. The surprisingly high German losses on day 1 are also remarkable, and the small losses on both sides prior to that are further evidence that some notion of battle "pace" needs to be acknowledged.

There are also other modeling questions that need to be dealt with:

- The losses listed in Table 1 are for the previous 24 h, so perhaps each day's losses should be based on the force levels of the previous day.

- Supplies of armor and artillery are crucial in such battles, so perhaps we should expand the database and include separate attrition coefficients for each distinct force component. There would be an attrition coefficient, for example, for Soviet artillery against German armor. Section 5.6 will describe an application involving many components on each side, with a correspondingly large number of attrition coefficients.
- Losses listed are "total" in the sense that they include noncombat losses. We would expect noncombat losses to be independent of the enemy force level, rather than being predicted by (say) the square law. Should we separate losses into combat and noncombat losses, with a separate forecasting formula for each?

Sheet "Kursk" of *Chapter5.xls* defines one possibility for dealing with some of the issues raised above. That sheet includes the data of Table 1, together with some functions needed to fit it. The losses between time t–1 and t ("losses on day t") are assumed to be related to the average of the force levels on day t–1 and day t. The adjustable parameters include a "pace" parameter for each day that measures the pace of battle relative to day 1. These parameters are multiplicative for the attrition on both sides on that day. There are 13 such pace parameters, plus two more for the attrition rates that define the square law. The idea is to adjust those 15 parameters to minimize some measure of error when forecast losses are compared to what actually happened in the battle.

On any given day, let the forecast losses be μ. According to the stochastic ideas of Section 5.3, the actual losses X should be a Poisson random variable with that mean. Since X is theoretically Poisson, its variance is also μ, so the quantity $(X-\mu)^2/\mu$ should be a random variable with mean 1. We can therefore make our measure of misfit be a sum of 28 such quantities, one for each side and day, and then adjust the 15 parameters to minimize this measure of misfit. If our assumptions about attrition are all correct, it should (when minimized) be on the order of about 28. On sheet "Kursk" of *Chapter5.xls*, this is accomplished using Excel's Solver.

The result of the minimization is that the two attrition parameters are 0.008 German casualties per Soviet man-day, and 0.038 Soviet casualties per German man-day (note the rough correspondence of these attrition rates to Engel's estimates for the battle of Iwo Jima). The minimized misfit is 5,581, much larger than the expected 28. The two worst error terms are due to the high German casualties on day 1 and the low German (compared to the Soviet) casualties on day 8, which together account for almost half of the misfit. Apparently the difference of a Poisson random variable from its mean is the least of our problems; the square law just does not fit very well, even after adjusting for the variable pace of battle. The reader is invited in Exercise 8 to find a better fit to the data in Table 1, but be warned: the Kursk data have been examined in detail by others (Turkes, 2000; Lucas and Turkes, 2004) without much success at finding a satisfactory fit.

Other battles of World War II have also been studied from a Lanchestrian perspective, e.g. Fricker (1998). The battle of Kursk is not alone in being hard to fit. A great deal of effort has been spent trying to fit simple attrition models to historical battles, much of it with similarly disappointing results. Part of the problem is that real battles are not designed well from a statistical standpoint. At Kursk, a significant problem is that German force levels remain nearly constant throughout the battle, and even Soviet force levels do not fluctuate very much. Battles to the finish (like Iwo Jima) are better, statistically speaking, but rare. Perhaps more significantly, historical battles are strongly influenced by subtleties that are hard to capture in a simple attrition formula and may have been understood only by the side initiating the action. Terrain is important in actuality, but missing from Lanchester's formulation. Logistics and morale are also usually missing, but frequently crucial in actual battles. If battle were as simple as Lanchester would have it, a general staff would hardly be necessary.

It does not follow from the frustrations outlined above that either Lanchester models or the statistical approach to estimating parameters is useless. Every study of Kursk has confirmed the strong lethality of the German troops involved, a tendency that was common to other battles involving the German army in World War II. The strong importance of numbers is confirmed by the battle of Iwo Jima. Given the multi-sided, surprise-laden nature of warfare, any combat model can be expected to err significantly when compared to reality. Models, however, approximate, are still a prerequisite for thought.

By the way, in spite of their strong advantage in terms of lethality, the outnumbered Germans eventually lost the battle of Kursk. It turned out to be the last German offensive on the Eastern front.

5.5 Aggregation and Valuation

Combat models are sometimes described as "aggregated" or "high resolution," but aggregation should really be measured on a continuous scale. Any combat model that counts things has made a judgment that two things are so similar that they can safely be thought of as identical. To the extent that dissimilar things are treated as if they were identical, the model is said to be more or less aggregated.

The Lanchester models introduced earlier in this chapter have counted only one kind of entity on each side and are therefore aggregated. There may be no harm in this aggregation, even though a closer look at the real world would reveal significant disparities among the entities. Suppose, for example, that we initially have a square law model where the attrition rates are a and b, and the initial numbers are x_0 for blue and y_0 for red. Suppose further that the blue forces are discovered to consist of half shooters and half cheerleaders. The cheerleaders perform some function that blue considers valuable, but do not shoot at red. If we were to leave the cheerleaders out of the battle, and simply double b, halve a, and halve x_0 to keep track only of the blue shooters, Lanchester's equations would predict the same survivors for red at all times. Since blue forces are reduced by half, but the lethality of the remainder is doubled, red attrition is unchanged (recall Equations 5.1). Since red's lethality coefficient is halved, so is attrition to blue, but then blue has only half the forces to attrit, so nothing essential about the battle changes. As long as we are careful about the meaning and measurement of lethality coefficients, it is a matter of convenience whether we count blue cheerleaders or not. Aggregation is not dangerous in this circumstance.

On the other hand, suppose red is able to distinguish blue cheerleaders from blue shooters and concentrates all of his fire on the blue shooters in the revised battle. This corresponds to doubling b, halving x_0, and *not* halving a (since all of red's fire is concentrated on the blue shooters). The revised battle now differs from the original; blue could very well win the original, but lose the revision (see Exercise 9). In this case, aggregation would be seriously misleading. Evidently, an important consideration in aggregation should be whether the other side can be expected to distinguish between the things being aggregated.

There is much more to be said about how much, if anything, is to be lost through aggregation (Davis, 1995), but sometimes there is no alternative. Engel (1954), for example, knew only the total count of Japanese troops initially on Iwo Jima. Constructing a model based on discriminating various types of Japanese soldiers was simply not an option.

In the authors' experience, the only method of aggregation used in practice is the weighted sum. If x_{ij} is the number of entities of type i and subtype j, then the equivalent number of aggregated type i entities is taken to be $x_i = \sum_j w_{ij} x_{ij}$, where w_{ij} is the "weight" of subtype j within the ith type. One of the weights is usually selected to be unity, so that x_i has the meaning "equivalent number of units of the selected subtype." Sometimes *all* of the weights are taken to be unity, reflecting the judgment that all subtypes are equivalent as far as combat is concerned. The Kursk data (CAA, 1998) include counts of tanks for both sides. There were actually a variety of tanks involved in that battle, so simply counting tanks is an example of using unity weights for all subtypes. The Kursk data include four fundamental components of combat power: combat manpower, armored personnel carriers (APCs), tanks, and artillery (only the manpower data are shown in Table 1). The discussion in Section 5.4.2 effectively uses the weights (1, 0, 0, 0) for the four components, since data about everything except manpower are ignored in that section. To construct a better scalar measure of total combat power, more realistic weights might be used. Aggregated Kursk models sometimes use the weights (1, 5, 20, 40) for the four components (Turkes, 2000), making the military judgment that having one additional tank is equivalent to having 20 additional men, etc. Whatever the source of the weights, the object in employing them is always to simplify through aggregation, even if some validity is lost in the process.

A weighted sum of components is an aggregated measure of total combat power. Consider two blue entities, each of which kills aggregated red combat power by killing a variety of red entities. If the first blue entity kills red combat power twice as fast as the second, then a good case can be made that the weight for the first blue entity should be twice that of the second. More generally, we might adopt the principle that a set of weights for the two sides is *consistent* if each entity's weight is proportional to the rate at which it kills the other side's weighted combat power. In a square law-type battle where all fire is "aimed," the consistency principle leads to a set of linear equations that can be solved for a nearly unique set of weights (Howes and Thrall, 1973). The essential mathematical problem in deriving these weights is that of finding the eigenvalue of a matrix, so the method is usually referred to as the eigenvalue method (there is also an ATCAL method, see CAA, 1983). The beauty of such methods is that they relieve the analyst of the need for making a priori judgments about military value, since the equations themselves determine proper weights.

In spite of its advantages, the consistency principle must be used with care. The eigenvalue method applies only to square law battles, and its logical basis becomes questionable if at any time some force component is reduced to zero. Also, the consistency principle would have the combat value of a truck being zero, since trucks do not actually shoot at anything. Logistic entities such as trucks are nonetheless frequent targets of enemy fire in the real world, and with good reason. Depending on purpose and circumstances, military judgment might provide a better set of weights.

5.6 The FAst THeater Model (FATHM)

The Department of Defense employs many models of combat in the defense planning process, one of which is FATHM. Although FATHM is a minor model in terms of usage, it has the advantage of having a simple structure that employs many of the features described earlier, so it makes a good case study. This section provides a concise description. Further details can in found in Brown and Washburn (2007).

FATHM is a deterministic model of large-scale air–ground combat. It is a simple model in the sense that it makes practically no reference to anything geographic – there is no front line, nor any reference to the range at which a weapon system is effective, nor any representation of terrain. It nonetheless incorporates considerable detail. FATHM includes data about a variety of platform types (tanks of various kinds, artillery, etc.) among its inputs, and its purpose is to determine the significance of these inputs. Three things happen in sequence in each of FATHM's three-day time steps:

- There is a battle between red and blue ground forces. This is the "ground" battle.
- There is a battle between the blue air force and the red ground force (blue control of the air is assumed, so there is no red air force). This is the "air" battle.

- Reinforcements arrive and a phase-change determination is made (wars are presumed to have phases, and the phase affects the nature of the ground battle).

Air and ground battles are intermixed in reality, but are imagined to be sequential in FATHM for practical reasons. Specifically, FATHM attempts to capitalize on the pre-existence of the COmbat SAmple GEnerator (COSAGE), a ground model, and the Conventional Forces Assessment Model (CFAM), an air model, integrating the two into an air–ground model that runs quickly on a computer. Speed is important on account of the anticipated use where different blue orders of battle are compared. The idea that air and ground are intermixed in time would be better modeled by shortening the time step, but this would cause FATHM to slow down. The choice of 3 days is a compromise between validity and efficiency.

The air and ground models each extensively employ EVA. In each battle in each time period, the number of surviving platforms is first calculated by a method that might be justified as being an accurate computation of the *expected* number of surviving platforms, but then this expected number is simply used as the actual number entering the next battle, even if it is not an integer.

5.6.1 FATHM's Ground Model

COSAGE is a high-resolution Monte Carlo simulation of ground combat, generally over a 2-day time interval, including all of the geographical features that are missing in FATHM. On account of its realism, it is far too slow to consider incorporating directly. Instead, FATHM approximates COSAGE with a deterministic Lanchester model. As far as FATHM is concerned, COSAGE is the reality that the Lanchester model must imitate.

COSAGE outputs include a "killer–victim" scoreboard that reports the total number of red platforms of each type killed by blue platforms of each type, and whether these kills are due to aimed (direct) or unaimed (indirect) fire. Indirect fire is usually artillery. Let I_{br} be the reported number of red platforms of type r killed indirectly by blue platforms of type b. Also suppose that the initial number of blue platforms of type b is B_{b0}, while the initial number of red platforms of type r is R_{r0}. If we equate COSAGE's indirect fire to Lanchester's linear law (equation (5.5)), blue's losses should be proportional to the number of red platforms and also to the number of blue platforms. Therefore, if $indir_{br}$ is the indirect fire coefficient, assuming that the COSAGE battle is 2 days long, we should have $indir_{br} \cong \dfrac{I_{br}}{(2\text{ days})B_{b0}R_{r0}}$. This formula calculates the attrition rate to be used in the Lanchester model as a function of quantities shown on the COSAGE killer-victim scoreboard. The reason for the "approximately equal" sign is that 2 days is not an infinitesimally small amount of time, as the Lanchester theory demands. We will return to this problem shortly, but, for the moment, assume that all of the indirect fire coefficients for blue against red are given by this formula. The red against blue indirect coefficients are given by the same formula with b and r reversed (the COSAGE output includes data about attrition to blue, as well as red).

Since direct red losses are proportional to blue force levels, but not to red force levels, the corresponding formula for the direct fire coefficient is $dir_{br} \cong \frac{D_{br}}{(2\text{ days})B_{b0}}$, where D_{br} is the number of red losses to direct blue fire according to the killer–victim scoreboard. Again, b and r can be interchanged to obtain the coefficients for direct red fire against blue.

With all of the coefficients determined, we can now make a Lanchester model of the battle. Letting B_{bt} and R_{rt} be the blue and red numbers of platforms of type r and b, respectively, at time t, we have $\frac{dR_{rt}}{dt} = \sum_b (indir_{br} B_{bt} R_{rt} + dir_{br} B_{bt})$, with the usual proviso that the equation holds only if $R_{rt} > 0$. Together with the corresponding equations for blue loss rates, these equations essentially constitute FATHM's ground model. There is a problem, however; if we were to solve these equations over a 2-day time period with the same initial forces as used in COSAGE, we would not measure the same losses. The reason for this is that Lanchester's equations are solved with an infinitesimal time increment, rather than a 2-day time increment. FATHM therefore "polishes" the raw attrition rates in an iterative scheme that ends when the Lanchester battle produces the same losses as the COSAGE battle. The polished rates seldom differ much from the raw ones. These polished rates are then employed throughout all time periods of FATHM's ground war, which may last for a 100 days or more. There is lots of arithmetic to do in the process, but ordinary differential equations are easy enough to solve numerically that the "fast" part of FATHM's name is still justified.

The COSAGE war is short enough that no platform type is completely wiped out, but FATHM's war is long enough that provision must be made for this possibility. Suppose that blue platform b is reduced to 0. No change in the indirect fire model is necessary, but the direct coefficient dir_{br} now represents the rate at which each red platform of type r aims lethally at a blue platform type that no longer exists. The idea is absurd, so the direct fire coefficients are modified in FATHM to proportionally reprogram this direct red fire to other blue target types. Similar modifications are made when red platform types are wiped out.

FATHM permits the initial number of platforms to differ from the number used in the COSAGE battle. Suppose that all FATHM inputs are double those of COSAGE. Direct attrition will also double, which is appropriate, but the indirect attrition will quadruple because it is proportional to both force levels. This quadrupling is actually desirable if all forces are in the same region as in the COSAGE battle, since the density of targets per unit area will double along with the rate of fire, but it is also possible that FATHM's user intends to model a battle over a larger region. For this reason, FATHM's user is required to input a parameter W that represents the frontal width of the battle region. For comparison, the COSAGE output also reports the width W_C of the COSAGE battle. The indirect attrition rates are then adjusted by multiplying them all by the ratio W_C/W. Thus, if the frontal width is doubled along with all of the platform numbers, indirect attrition will only double.

In summary, FATHM's ground model is a Lanchester model that takes advantage of the data available in the COSAGE killer–victim scoreboards, while still being able to replicate the results of the COSAGE battle. Adjustments to coefficients are made to account for the different time and space scales of the typical FATHM battle.

As mentioned earlier, ground battles have phases that depend on the defensive/offensive nature of the two sides. There is a separate COSAGE run for each phase, along with separate sets of implied attrition coefficients. When the right criteria have been met, FATHM changes phases by substituting the appropriate set of attrition coefficients.

5.6.2 FATHM's Air Model

FATHM's air battle amounts to an allocation of blue aircraft sorties to red targets (enemy platform types such as T-72 tanks). The targets shoot back only in the sense that the database recognizes aircraft attrition probabilities that depend on the target type.

The allocation of aircraft to targets is optimized using linear programming (Appendix B), in spite of some relationships that are fundamentally nonlinear in the real world. For example, suppose a given target is attacked by n sorties, each of which has a kill probability of P_K. If the sorties are independent, then "powering up" would lead to a kill probability of $1-(1-P_K)^n$, a nonlinear function of n. FATHM instead uses the linear formula nP_K, adding a constraint that the overall kill probability cannot exceed 1. This is equivalent to using the upper bound discussed in Section 3.4.2 to approximate the number of targets killed. Thus, if the kill probability is 2/3, at most 1.5 sorties can be assigned to the target, and the target will surely be killed if that number of sorties is assigned to it.

Forcing the mathematical program to be linear is dictated by FATHM's ambition to be fast. A linear program is the simplest type to solve, and even solving the linear program is a significant task because of the large variable count. FATHM's linear program may have on the order of 80,000 variables! Why are there so many? Suppose that each of 100 target types could be attacked in one of 100 ways. Then there would have to be 10,000 variables, since each doubly subscripted variable would stand for the number of targets of a certain type attacked in a certain way. FATHM's allocation variables actually have six subscripts, so the large variable count should not be surprising. There are also a similarly large number of constraints. The primary constraint is that the number of sorties cannot exceed whatever amount can be generated in 3 days, but there are also constraints associated with the subsidiary goals of killing enough red targets of certain specified types to permit the phase of the war to advance, and more constraints in addition. Solving even one linear program of this size is not trivial, and a separate linear program is required for the air battle in each 3-day time segment. Solution of a nonlinear program in those circumstances would be too time consuming.

The object of the air battle is to kill as much red target value as possible, so red targets must be valued. FATHM's method of doing this is a hybrid between military judgment and the consistency principle of Section 5.5. All platforms on both sides are first assigned a static value. One frequently used scale for this static value is replacement cost in dollars – military value should not differ much from replacement cost because the replacement process is itself budget constrained. In each 3-day time period, each surviving red platform will kill a certain amount of blue static value. The dynamic value of that red platform is taken to be that amount of blue static value, but multiplied by a constant with units of time that represents the time horizon. The total value of the red platform is then the sum of its static and dynamic values. In this way the ground model communicates to the air model which red targets are particularly crucial to kill. The total value of a red truck is simply its static value, but the total value of a red tank will also include a term proportional to the rate at which that tank is currently killing blue static value. It is not necessary to find a similar total value for blue platforms; since FATHM is a one-sided model in terms of decision making.

5.6.3 Implementation and Usage

FATHM is complied FORTRAN code that exploits the capabilities of Excel™ workbooks for input and output. The user adjusts the force levels, static target values, etc., in a multi-sheet workbook, and then outputs comma-delimited files for later use by FATHM. FATHM reads these files, together with the killer–victim scoreboards from prior COSAGE runs, computes the Lanchester attrition coefficients as described above, fights the multi-phase war, and then outputs the overall results about attrition to another comma-delimited file. This output file can then be input to Excel™ and displayed or manipulated as the user wishes. Figure 8 is a sample graphical output showing which blue platforms are causing the most attrition. Data can be displayed in other ways using Excel's Pivot tables.

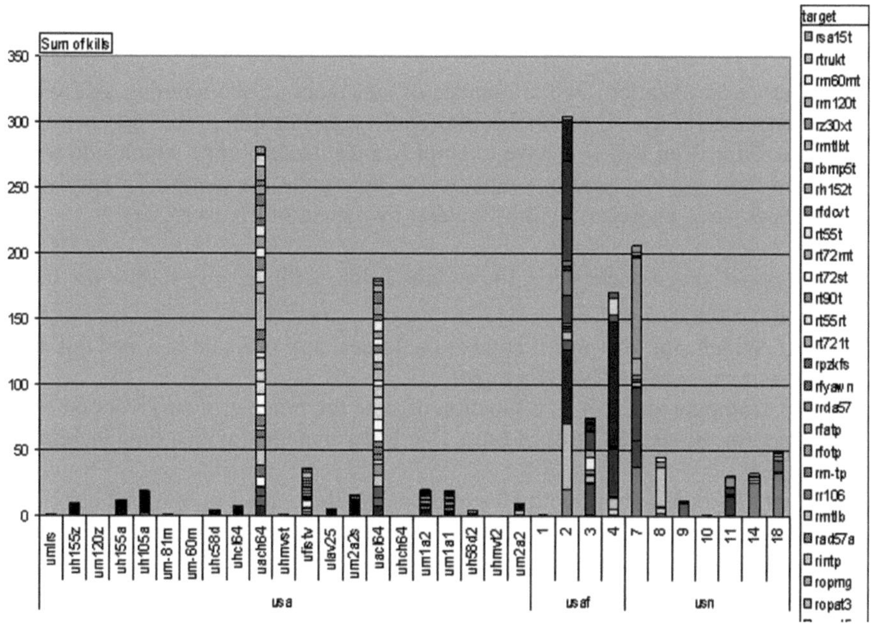

Figure 8: Attrition results of a FATHM battle. Each segmented vertical bar shows *red* platforms killed by each type of *blue* platform, with the segments corresponding to the code shown on the right (originally in color). Shown are 21 army (usa) platforms, 4 air force (usaf) platforms, and 7 navy (usn) platforms. All results (including the platform counts) are fictitious.

Exercises

(1) Derive an equation for the number of survivors on the winning side when Lanchester's square law holds and battle continues until one side is wiped out. Hint: You will first have to employ a test to determine which side wins, and then solve a quadratic equation to determine the number of survivors. Check your answer using the "SquareLaw" sheet of *Chapter5.xls*.

(2) Suppose that Lanchester's linear law holds, with $x_0 = y_0 = 100$, $a = 0.01$, and $b = 0.02$.
(a) Which side will win if combat continues until one side is wiped out, and how many survivors will there be?
(b) Compute and plot as a function of time the number of survivors on each side, for the first 50 units of time. Use Euler's method with a time increment of 0.5.
Ans. In (a), blue wins with 50 survivors. Neither side is ever wiped out completely when the linear law holds, so the two curves in (b) should gradually approach the horizontal axis, rather than dive through it. Make sure that your answer to (b) is consistent with (a).

(3) Lanchester (1916), by way of emphasizing the importance of concentration of forces, considers a square law example where $a = b = 1$ and both sides initially have 1000 units. If a single battle occurs, then both sides run out at the same time. What happens if blue divides his force in two, so that there is a first battle with 500 blue units and 1000 red units, followed by a second battle with 500 blue units and the red survivors of the first battle?
Ans. Red wins with 707 survivors.

(4) Let $a = b = 0.1$, $\Delta = 1$, $x_0 = y_0 = 1000$, and consider the solution of the difference equations (5.2). This is a special case of Hughes' salvo equations where the two sides are exactly matched. If (5.2) is solved as given, the two sides run out at the same time. What happens if red strikes first in each cycle; that is, what happens if the update formula for $y(t + \Delta)$ replaces $x(t)$ by $x(t + \Delta)$? How much of an advantage is this for red? Quantify the answer by using a spreadsheet to find the number of red survivors. To be precise, find the number of red survivors after the last cycle in which the number of blue survivors is still positive.
Ans. 316 red survivors.

(5) In Example 5, the "Endgame" sheet of *Chapter5.xls* was used to find the average net difference of blue survivors (3.38). Use the same sheet to find the average number of blue survivors and the average number of red survivors. Is the difference between them 3.38?

Ans. 4.60 blue survivors and 1.22 red survivors, so the difference is indeed 3.38.

(6) Example 6 refers to sheet "Tanks" of *Chapter5.xls*. The initial number of blue tanks in that battle is 20, but change that initial cell to reflect the idea that you can have more tanks if you are willing to adopt a cheaper command system that takes a longer time to shoot after a miss (larger τ_m). Specifically, let the initial number of tanks be $19+\tau_m$. For $0.5 \leq \tau_m \leq 2$, what is the best value of τ_m?
Ans. There are 7.7 blue survivors if $\tau_m = 0.5$, and smaller numbers otherwise. In this case 19.5 high-quality tanks is better than a larger number of tanks that take longer to shoot.

(7) Sheet "Iwo Jima" of *Chapter5.xls* replicates the battle of Iwo Jima using Engel's estimated attrition rates. Experiment with the initial number of Japanese troops. What is the smallest initial number that would have left the US with zero active troops after 60 days of fighting? Excel's Tools/Goal Seek function may be of use.
Ans. 34,732.

(8) Sheet "Kursk" of *Chapter5.xls* is set up to minimize the measure of misfit described in Example 7. See if you can find a better fit than the one described there, without using too many parameters, or a slightly worse fit that uses fewer parameters. You might, for example, try the linear law or even the logarithmic law (where attrition is proportional to own force level, not the enemy's). What if all the pace parameters are forced to be 1? Does the fit suffer much?

(9) Compare two square law battles between blue (x) and red (y). In the original battle, $x_0 = 200$, $y_0 = 300$, $a = 0.04$, and $b = 0.01$. In the revised battle, $x_0 = 100$, $y_0 = 300$, $a = 0.04$, and $b = 0.02$. The idea in the revised battle is that red has figured out that blue's 200 units include only 100 that actually shoot and concentrates all his fire on those shooters. Compare the two battles.
Ans. In a battle to the finish, blue would win the original battle with 132 survivors. Red would win the revised battle.

(10) The y side has 1000 men in each of several forts, with $a = 0.02$. The x side has 4100 men in a single group, with $b = 0.01$. The x side attacks the forts in succession, with the survivors of one battle entering the next. How many forts are taken before x is wiped out, assuming each battle proceeds until one side is exhausted? Assume first that the square law holds and next that the linear law holds.
Ans. 8 for the square law, 2 for the linear law.

(11) The equations $dx/dt = -k_1$ and $dy/dt = -k_2$ for $x, y > 0$ could be said to hold in "gladiator" combat where there are successive one-on-one combats. Compare the deterministic and probabilistic solutions when $k_1 = k_2 = 1$ and $(x_0, y_0) = (3,2)$.

(12) Consider a Square law battle where $(x_0, y_0, a, b) = (6,5,0.8,0.9)$, interpreted probabilistically. Compute the average net difference of survivors $x - y$ at the end of the battle. Use the "Endgame" sheet of *Chapter5.xls* to calculate the difference analytically, and also use the "SquareLawRnd" sheet together with *SimSheet.xls* to compute it by Monte Carlo simulation. The results will not be exactly the same, but should be close.

(13) In the Battle of Ein-A-Tinna (Lebanon War 1982), the theater of operations was southern Lebanon, which is a mountainous region crisscrossed by narrow, steep, and winding roads. Mechanized units were typically forced to move in a single column. An Israeli tank battalion approached the village of Ein-A-Tinna from the south assuming that there were no Syrian troops in this village. As the first tank in the column turned around a horseshoe bend in the road, it was disabled by heavy fire from the village. The second tank in the column passed it and took over the position in the front, resuming the duel. It was also disabled, as was a third tank when it moved to the front. Later on the combat situation became more complex when another Israeli force approaching the village from the Northeast misinterpreted the situation and initiated a fire exchange with the Israeli forces in the south, causing some heavy fratricide casualties. As far as the first part of the battle is concerned, a small Syrian force (~6 tanks) managed to stop a battalion because it was able to exercise all of its fire power while the firepower of the Israeli forces was reduced, due to the topography, to a single tank.
To make a Lanchester model of this situation, let m and n be the numbers of surviving Israeli and Syrian tanks, and assume that all Syrian tanks can fire at Israeli tanks, while only one (the lead) Israeli tank can return fire. Also assume that all tanks are equally effective, and that the battle continues until the loser is annihilated. Except for the restriction on Israeli fire, Lanchester's square law would apply. If there are initially two tanks on each side, what is the probability that Israel wins? Also, does it matter if some Syrian fire is directed at Israeli tanks that are not in the lead?
Ans. Since the length of the battle is not in question, use the Markov formula (5.10) with $Q(m,n) = n/(1+n)$. If $MOE(m,n)$ is the probability that Israel wins, then $MOE(m, 0) = 1$ for $m > 0$, while $MOE(0, n) = 0$ for $n > 0$. After applying (5.10) a few times, the answer is $MOE(2, 2) = 0.36$. Sheet "Endgame" of *Chapter5.xls* can be used to do this if the correct transition probabilities are substituted for the function called "leftprob" on that sheet. It does not matter how the Syrian tanks direct their fire, as long as all fire is aimed at Israeli tanks.

(14) The analytic solution of equations (5.1) over the time period when both state variables are positive is

$$x(t) = \frac{1}{2}\left(x_0 - y_0\sqrt{\frac{a}{b}} \exp(t\sqrt{ab}) \right) + \frac{1}{2}\left(x_0 + y_0\sqrt{\frac{a}{b}} \exp(-t\sqrt{ab}) \right)$$

$$y(t) = \frac{1}{2}\left(y_0 - x_0\sqrt{\frac{b}{a}} \exp(t\sqrt{ab}) \right) + \frac{1}{2}\left(y_0 + x_0\sqrt{\frac{b}{a}} \exp(-t\sqrt{ab}) \right)$$

Here x_0 and y_0 represent the given starting values $x(0)$ and $y(0)$. Show by differentiation that (5.1) is indeed satisfied by these analytic equations.

Chapter 6
Game Theory and Wargames

The game's afoot!

Sherlock Holmes

6.1 Introduction

Military operations are conducted in the presence of uncertainty, much of which is due to unpredictable enemy actions. The part due to the enemy is a different kind of uncertainty than not knowing, for example, whether a rocket will hit its target. In the case of the rocket, we are dealing with a device that itself has no feelings or desires about the outcome. There are many sources of that kind of randomness in both military and civilian affairs, all of which we attribute here to choices by "Mother Nature," a fictional decision maker who makes choices unpredictably, but never with any regard for their consequences. We often quantify Mother Nature's actions by providing a probability distribution over her choices, as in all previous chapters of this book.

Uncertainty about enemy action deserves a different kind of treatment, since an enemy by definition has strong feelings about the combat outcome. Decision making against an enemy differs qualitatively from decision making against Mother Nature. Ships traveling through submarine-infested waters sometimes adopt a randomized zigzag course, the idea being to make it difficult for a submarine to aim a torpedo accurately. Ships traveling through waters infested with icebergs are also at risk, but they do not zigzag, nor do they attempt to behave randomly in any manner. Both situations involve uncertainty, but there is something fundamentally different about the actions that are called for when dealing with a sentient enemy.

Analysts have followed two separate lines of thought in trying to deal with multi-sided decision-making situations. One line of thought is theoretical and leads to game theory (Section 6.2). The other is experimental and leads to wargaming (Section 6.3). Both lines of thought will be seen to have advantages and disadvantages. The disadvantages are significant, and partially explain why most combat models are actually single sided. The advantages should not be lost sight of.

One advantage of the multi-sided approach has to do with the use of combat models to evaluate the effects of new tactics or equipment. Single-sided combat models are forced to rely on more or less sophisticated decision rules for all participants other than the innovating side. These rules typically attempt to imitate past behavior of opponents. If the new tactic or equipment is a significant departure from the past, the effectiveness of the innovation may be exaggerated in a

A. Washburn, M. Kress, *Combat Modeling*, International Series in Operations Research & Management Science 134, DOI 10.1007/978-1-4419-0790-5_6,

single-sided model through lack of sufficient adaptability in these rules. The modern machine gun would have been devastating against the kind of massed infantry charges that occurred in the American Civil War. The ultimate effect of the machine gun was to change the nature of infantry tactics to mitigate its effects, but it would have been hard to anticipate this change in the rules of a civil war era single-sided model. Multi-sided models have a better chance of anticipating the appropriate tactical reaction to such innovations, either through the inherent cleverness of humans (in wargaming) or through optimization (in game theory).

6.2 Game Theory

Do not be misled by the name – the theory is not aimed merely at parlor games such as checkers or rock-paper-scissors, although we will often use such games as examples. The whole topic was invented at the end of World War II by von Neumann and Morgenstern (1944), who make it plain in their seminal book that their ambition is to apply game theory to military and economic situations, as well as parlor games. The central feature of all these situations is that the game's outcome is determined by the actions of multiple decision makers, and that they do not all desire the same outcome. When formalized by specifying each player's utility as a function of all of the players' strategic choices, we have what von Neumann and Morgenstern call a game.

Two-person zero-sum (TPZS) games are a special class of game with exactly two players who are completely opposed, as in most parlor games. Chess, backgammon, tic-tac-toe, and rock-paper-scissors are examples. So is baseball – even though each side or "player" in baseball is composed of several humans, the important thing is that all of the humans on each side are in complete agreement as to goals. Monopoly (when played by more than two players) is not TPZS.

The rest of this chapter is devoted entirely to TPZS games. There are two reasons for this restriction:

- combat usually involves two opposed sides
- the TPZS theory is in better shape than the theory of games in general

With regard to the second reason, the curious reader should search the Internet for "Prisoner's dilemma" as an example of a fascinating, two-person nonzero-sum game that has been and remains the object of much analysis. In spite of its simplicity and large amounts of pertinent thought, there is still no commonly accepted "optimal" way to play the game. We do not mean to deny the importance of general games, which are being used increasingly to model economic situations such as auctions, and which do potentially have military applications. However, the TPZS part of game theory will suffice for our purposes here.

6.2.1 Matrix Games and Decision Theory

The rules of any game must specify the possible actions or "strategies" available to each side, and, as a function of these actions, the payoff to each player. In the case of TPZS games a single payoff function suffices, since the players are completely opposed. It is conventional to define player 1 as the maximizing player and player 2 as the minimizing player, in which case the payoff function is essentially player 1's utility for the games's outcome. If each player has only a finite number of strategies, the rules of the game can be shown by giving a single matrix that represents player 1's payoff, with one row for each of player 1's strategies, and one column for each of player 2's strategies. The entry in the *i*th row and *j*th column is usually called a_{ij}. This is the payoff that player 1 wants to maximize, while at the same time player 2 wants to minimize it.

It happens that the same matrix notation can also be used to represent decision problems where player 2 is replaced by Mother Nature. Before considering such matrices as games, let us review the conventional decision theoretic solution. Mother Nature chooses the column, and the decision maker chooses the row without knowing her choice. To quantify Mother Nature's actions, we introduce a probability distribution **y** for which y_j is the probability that Mother Nature chooses column j. The average payoff when row i is chosen by the decision maker is then $avg_i \equiv \sum_{j=1}^{n} y_j a_{ij}$, where n is the number of possible choices by Mother Nature. Player 1 should select i to maximize avg_i, his expected utility.

Example 1: In World War II, there was a time when aircraft took off carrying depth charges that they hoped to use in sinking submarines. Each depth charge could be set before takeoff to explode at either a shallow or deep depth. The submarine target would also be either shallow or deep, depending on how much warning the submarine got of the approaching attack. We wish to model this situation as a decision problem where the aircraft decides how to set the depth charge. In setting the depth charge, the aircraft does not know the depth of the submarine that it will be used against, but it does know the payoffs shown in Table 1, which displays the conditional probability of sinking the submarine as a function of the depth charge setting (sh or dp) and the submarine depth (SH or DP).

In Table 1, the best setting against SH is sh and the best setting against DP is dp, but the aircraft does not know whether SH or DP will be the case when setting the depth charge. The conventional solution to this dilemma is to introduce a distribution **y** on SH or DP, which we take to be $\mathbf{y} = (0.4, 0.6)$. Thus, while most submarines have managed to be DP by the time the aircraft actually drops the depth charge, 40% are still shallow (SH) at the time of the attack. In World War II, that particular distribution was based on operational experience. Since $avg_{\text{sh}} = 0.4 \times 0.5 + 0.6 \times 0.1 = 0.26$ and $avg_{\text{dp}} = 0.17$, the best depth change setting is sh. The two avg numbers represent unconditional sink probabilities according to the theorem of total probability, so the aircraft should expect success in 26% of its attacks if it sets the depth changes to sh. Three things are true in this example:

1. most submarines are DP when encountered,
2. the best depth charge setting for DP submarines is dp, and
3. the best depth charge setting is sh.

This example is actually a true story, except for the assumed precision of the numbers. It is often told to illustrate the advantages of a quantitative approach based on decision theory (Morse and Kimball, 1950, their Section 3.4.8). In the press of combat, it is easy to make the first two observations, and come to the "logical" conclusion that the best setting is dp. A more careful analysis leads to the opposite conclusion and a significant increase in effectiveness.

	Shallow submarine (SH)	**Deep submarine (DP)**
Shallow depth (sh)	0.5	0.1
Deep depth (dp)	0.125	0.2

Table 1: Conditional probability of sinking the submarine in four cases.

The assumption that **y** = (0.4, 0.6) is a crucial part of the analysis in Example 1. German submarines of the era routinely reacted to the sighting of an aircraft by submerging, but were occasionally surprised on the surface and ended up at a shallow depth when attacked. Thus the SH fraction is essentially "fraction of the time surprised," over which the submarines had no control. Suppose we take away that difficulty, which was peculiar to the equipment of the era, give the submarine commander a free choice as to depth, and ask again what the depth charge setting should be. We can no longer assume **y** = (0.4, 0.6), but it is still necessary to make some kind of assumption about the submarine's choice.

One such assumption is the "conservative" assumption of worst-case analysis. This amounts to assuming that the submarine will be DP if the setting is sh, or SH if the setting is dp. Our conclusion would then be that the best setting is dp, and that the submarine will be sunk 0.125 of the time (the sh setting would lead to only a 0.1 sink probability). Regardless of tactics, the submarine will be sunk at least 0.125 of the time if depth charges are consistently set to dp.

Worst-case analysis makes two assumptions about player 2's behavior. One is that he is opposed to player 1. The other assumption is omniscience, since player 2 is in effect assumed to select a column in full knowledge of player 1's row choice. The game theory idea is to retain the first assumption, but not the second. Certainly the German submarines of World War II were not omniscient – how could the submarine commander know the depth charge setting? Just as the aircraft does not know the submarine depth, neither does the submarine know the depth charge setting. The two players are treated symmetrically in game theory, with neither knowing the choice of the other.

The trouble is that the assumption that each player acts in ignorance of the other's choice seems at first to go nowhere. The aircraft, thinking that the submarine prefers the DP setting, might set the depth charge to dp. But then the submarine, thinking that the aircraft might think that the submarine will choose DP, might actually choose SH. This kind of "if he thinks that I think that he thinks…" regression clearly has no end, and cannot be the basis for a theory of action. The TPZS theory is not based on it.

The solution for both sides turns out to be to simply act randomly, albeit with a carefully selected kind of randomness. von Neumann (1928) shows that all finite TPZS games have a satisfactory notion of solution if such randomized or "mixed" strategies are employed. Here is what we mean by "satisfactory":

Definition: Given the payoff matrix (a_{ij}), an $m \times n$ TPZS game has a solution if there exist probability distributions **x** and **y** and a number v such that $\sum_{i=1}^{m} x_i a_{ij} \geq v$ for $j = 1,\ldots,n$ and $\sum_{j=1}^{n} y_j a_{ij} \leq v$ for $i = 1,\ldots,m$. If all conditions are satisfied, the probability distributions are called "optimal mixed strategies," the number v is called the "value of the game," and (**x**,**y**,v) is the solution of the game.

The first inequality condition in the definition states that, if player 1 selects a row according to **x**, the average payoff will be at least v, no matter what player 2 does, even if player 2 knows **x**. The second condition guarantees that the payoff will never exceed v if player 2 acts according to **y**, even if player 1 knows **y**. Taken together, the two conditions establish that v can be guaranteed by each side, regardless of the actions of the other, and therefore deserves to be called the game's value. Do not be put off by the word "average" here. There is no implication that the game must be played repeatedly. Even if you only gamble once, you might as well take the gamble with the largest probability of winning.

Our depth charge game has a solution, and its value is 7/38 (approximately 0.18). Player 2 can guarantee that the sink probability will not exceed that number by using **y** = (4/19, 15/19). To confirm that, use that distribution for Mother Nature's choice in a conventional decision theory problem (Exercise 1). You should find that player 1 has two choices, both of which lead to a sink probability of 7/38. In fact, the game theory solution can be thought of as worst-case analysis where a "case" is a probability distribution for Mother Nature – there is no worse distribution than **y**. That same game value can be guaranteed by the aircraft using **x**=(3/19, 16/19). If each depth charge is set to be sh or dp according to that distribution, there is nothing the submarine can do to prevent being sunk 7/38 of the time. Even if **x** is implemented by including a random number generator in the depth charge itself, so that no action by the pilot is necessary, and even if one of those depth charges is captured and exploited by the submarine force, there is still no way to prevent being sunk 7/38 of the time. Neither party needs to keep his optimal mixed strategy secret, although the choice from that distribution on any particular play of the game should of course not be revealed. Both mixed strategies have the property of equalizing the average payoff as seen by the other player, and each can be derived by writing two linear equations to that effect (Exercise 2).

All finite TPZS games have a solution, as von Neumann (1928) has shown. His proof is not constructive, but by now there are efficient ways of computing solutions. One way is to solve an optimization problem where one of the players optimizes the variable v, subject to appropriate constraints from the definition above. For example, the problem for player 2 is to minimize v, subject to the constraints that $\mathbf{y}$ must be a probability distribution (sums to 1, no negative numbers), and $\sum_{j=1}^{n} y_j a_{ij} \le v$ for all i. Sheet "GameSolve" of *Chapter6.xls* has an implementation where the user inputs an arbitrary game matrix and Solver finds the solution, including the optimal $\mathbf{y}$. The optimal $\mathbf{x}$ could be determined from a similar optimization for player 1 and is guaranteed to produce the same game value. However, player 1's mixed strategy turns out to be the dual variables in player 2's minimization, so a single minimization does it all. Sheet "GameSolve" exploits this, reporting both players' optimal strategies after a single minimization.

von Neumann's existence theorem stipulates that the number of strategies must be finite. Some games with infinitely many possible strategies also have solutions, but some do not. An example of the latter is "pick the largest number." Two players each secretly write down a number on a piece of paper. The numbers are then compared, and player 1 wins if and only if his number is bigger than player 2's. The rules are perfectly well defined, but there is clearly no optimal way to play that game, with or without a random number generator.

This section has so far summarized about a century's thought on how to deal with competitive decision-making problems. The results were by no means obvious before 1928 – von Neumann's theorem was once thought to be false by some powerful mathematicians. Nonetheless, the fact is that all finite TPZS games are now known to have a solution if mixed strategies are permitted. You may find this triumph unimpressive. Most people are not attracted to the idea of making decisions by randomization, whatever the seeming logical necessity. It is the authors' experience that military officers, in particular, are repelled by the idea of responding to questions like "Captain, should we turn right or left?" with "Flip a coin!" The idea of basing action on a guess at the opponent's choice is more satisfying, especially, of course, if the guess turns out to be correct. While von Neumann has shown that it is not necessary and possibly unwise to go through the process of trying to outguess your opponent, the fact is that most people would like to try it. Evidence of this is the existence of the World RPS Society, which stages tournaments where experts at rock-paper-scissors compete against one another. RPS is easily shown to be a TPZS game where the optimal strategy for either side is to select each of the options with probability 1/3 (Exercise 3), so any neophyte with a random number generator can avoid losing to the world's greatest RPS player by simply choosing strategies at random. The society's members surely know that, but still enjoy playing the game.

In spite of our human predilection for attempting to outguess the opponent, von Neumann's theorem still stands. If nothing else, at least take note that the game value v is a way of summarizing a complicated situation with a single equivalent number. In a combat model, we might say that every meeting between an aircraft and a submarine will sink the submarine with probability 7/38, never mind the details, as long as both players are rational. Such simplifications are often valuable.

Example 2: Colonel Blotto has three regiments available for attacking two forts, and must decide how many regiments to use on each fort. His opponent has four regiments available for defense, which he must also divide between the two forts. Each party knows the overall strength of the other, but neither party knows how the other's forces will be divided. Blotto takes a fort if and only if his assignment is strictly greater than the defender's, and his goal is to take as many forts as possible. How should he divide his forces?

We first note that worst-case analysis will result in Blotto taking no forts at all, since he is outnumbered. Blotto's hope lies in the defender's not being able to predict how Blotto will divide his forces.

To solve this problem as a game, we first identify all of the strategies that are feasible for Blotto and his opponent. Blotto has four strategies and his opponent has five. Figure 1 shows the payoff matrix, using the notation that bold **uv** is the strategy of using u units on the first fort and v on the second (this usage is an exception to our usual policy that bold symbols stand for vectors or matrices).

After providing the "Gamesolve" sheet with a 4×5 game matrix (the row and column names are not inputs) and solving the game, we find that a solution is $\mathbf{x} = (0.5, 0, 0.5, 0)$, $\mathbf{y} = (0, 0.5, 0, 0.5, 0)$, and v=0.5. Blotto can guarantee to win one fort half the time by flipping a coin to decide between **30** and **12**, but cannot guarantee anything more than that. An alternative optimal mixed strategy for Blotto is to flip a coin between **30** and **03**. The defender flips a coin to choose **31** or **13**.

	40	**31**	**22**	**13**	**04**
30	0	0	1	1	1
21	1	0	0	1	1
12	1	1	0	0	1
03	1	1	1	0	0

Figure 1: Showing the number of forts won by Blotto as a function of his own division of forces and his enemy's. Strategy names are in *bold.*

The power of modern computers is such that games with thousands of strategies can be routinely solved. The depth charge game, for example, would be solvable even if both sides had thousands of different choices for depth, rather than only two.

In spite of the efficiency of modern linear programming codes, however, many games are simply too big to solve in the matrix form described above. If Blotto had 10 regiments to attack 10 forts, the payoff matrix would have 92,378 rows, rather than 4, and the number of rows would be still larger if the 10 regiments were not identical. As a practical matter, even listing all of Blotto's possible strategies would be a daunting task. Although any finite game can be solved *in principle* by first constructing its payoff matrix, it will sometimes be advisable to represent the game in a different manner. The following sections deal with some of the possibilities.

6.2.2 Tree Games and Saddle Points

Many games involve a sequence of moves, with each move being a choice for player 1, player 2, or Mother Nature (chance moves). The rules of such games may be better expressed in the form of a tree, rather than a matrix. Figure 2 shows an example. Play begins at the root of the tree, where player 1 must choose between three branches, two of which lead to a choice for player 2, and one for Mother Nature. Payoffs are shown at the terminal "leaves" of the tree. Player 1 maximizes, player 2 minimizes, and Mother Nature randomizes according to probabilities shown on the branches for each of her decisions (solid dots in Figure 2).

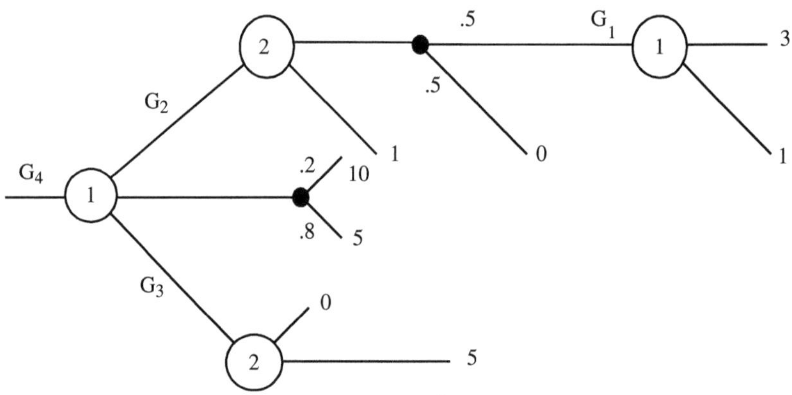

Figure 2: A simple tree game. ***Circles*** **are moves for the player whose number appears within the circle.** ***Solid dots*** **are randomizations, with probabilities given on the branches. Play begins at the root and ends at a leaf where further branching is impossible. The number given at each leaf is the payoff. The G_n symbols are for discussion only and are not part of the rules: G_n refers to the subtree that follows the symbol and G_4 is the root.**

Although play begins at the root, the best way to find optimal strategies is by beginning computations at the end and doing a "backward rollup." Thus, at point G_1, player 1 will choose 3 rather than 1, so point G_1 can be thought of as a leaf where the payoff is 3. The preceding move for Mother Nature is therefore worth the average of 3 and 0, which is 1.5. Since player 2 prefers 1 to 1.5, point G_2 is equivalent to a leaf whose value is 1. The value of G_3 is 0, so the value of the game G_4 is the larger of 1, 6, and 0, which is 6. This logic is automated on sheet "Tree" of *Chapter6.xls*. When the game is played optimally by both sides, player 1 selects the middle branch at G_4, Mother Nature chooses either 10 or 5 according to the given probabilities, and the game ends. In effect, G_n is a subgame whose payoffs may include other games.

The tree game shown in Figure 2 is boring when played (player 2 never even gets to choose anything), but still exhibits all of the features of the type. Tree games include more interesting games such as tic-tac-toe and chess (which have no chance moves), backgammon (which does), and most military board games that involve the rolling of dice and a playing board that is at all times visible to both sides. Chess, for example, has an initial move for White with 20 branches (16 pawn moves and 4 knight moves), and eventually payoffs of 1 (White wins), −1 (Black wins), or 0 (draw) at the leaves. Furthermore, all such games can be solved in principle by the backward substitution technique. The values of tic-tac-toe and checkers are both known to be 0. The value of chess is unknown but is surely either 1, 0, or −1. The value of backgammon is 0, by symmetry, but the optimal strategies for playing the game are unknown. It is only in the case of tic-tac-toe where the method of explicitly drawing the tree and using backward substitution might actually be practical – the tree in the other cases is far too large to deal with. Nonetheless, representing the game as a tree is still a useful conceptual device.

Tree games can also be represented as matrix games. One of player 1's strategies for the game of Figure 2, for example, is "first move to G_2, and then take the 1 if we ever get to G_4." That strategy is not very smart because of its behavior at G_4, but all we require of any strategy is that it be a complete rule for choice no matter what the moves of the other player. Our initial goal is to simply list them all, rather than to seek strategies that are optimal. Player 1 has three other distinct strategies, and player 2 also happens to have four strategies. One of player 2's is "take the 1 at G_2 or take the 5 at G_3." If player 1's named strategy is matched with player 2's, the payoff is 1. With sufficient patience, we could name all four strategies for each side and fill out the entire 4×4 payoff matrix for this game. If we do that, we will find a matrix element that is simultaneously the largest entry in its column and the smallest entry in its row, i.e., a saddle point of the matrix. For the game in Figure 2, the row involved will be "at G_4, choose to hand it over to Mother Nature," and the value of the saddle point will be 6.

If a payoff matrix has a saddle point at (i,j), then player 1 can guarantee a_{ij} by choosing row i, regardless of what player 2 does, and player 2 can guarantee a_{ij} by choosing row j, regardless of what player 1 does. It is not necessary to employ mixed strategies in such games, since the game value can be guaranteed by each side without resort to randomization. Tree games are sometimes called *strictly determined* for this reason. There is thus no computational point in representing a tree game as a matrix game; nothing new is discovered, and backward substitution is a more efficient solution technique anyway. The point here is that tree games are simply a special kind of game for which we have a more efficient solution method than linear programming.

Tree games are also sometimes called games of *perfect information*, the essential point being that, whenever it comes time for a player to make a choice in a tree game, he knows exactly where he is in the tree. An adequate theory for finite tree games (Zermelo, 1912) existed well before 1928, when von Neumann proved his theorem for all finite TPZS games.

While many parlor games are games of perfect information, there are also many that are not, and which therefore cannot be represented as trees. These include most card games (cannot see the other player's hand), blind chess (cannot see the other player's pieces), battleship (cannot see where the other player has located his ships), and rock-paper-scissors (cannot see what the other player's symbol is, at least not in time to affect one's own choice). An attempt to represent rock-paper-scissors as a tree game would have to put either player 1 or player 2 at the root, and that player would lose.

6.2.3 Solvable Games

We have presented two methods for solving games, linear programming for the matrix games of Section 6.2.1, and back-substitution for the tree games of Section 6.2.2. However, there are lots of interesting, large TPZS games that are ill-suited to either format. Although there have been many solutions of large TPZS games, most involve the exploitation of some special feature that makes it unnecessary to begin by listing all possible strategies, as in matrix games, or by completely laying out a game tree. Without being specific about exactly how this exploitation occurs, the object of this section is to describe some of the classes of large TPZS games for which solutions are computable as a practical matter.

Guessing Games. One of the most common direct sources of matrix games is where player 2 chooses a number and player 1 is rewarded in some way for guessing approximately the same thing. The number might be the depth of a submarine, as in the depth charge example given above. It could also be a location at which to hide, a frequency on which to communicate, a time at which to act, or a road on which to travel. The payoff is usually highest when player 1 is nearly correct. The value of the game is sometimes computed and used for subsequent investigations. For example, an equivalent detection range might be computed as the value of a game in which a submarine chooses its own depth and a surface ship chooses the depth of a dipping sonar (see Exercise 5). Section 8.5.2 is an example of a kind of guessing game that arises in mine warfare. Washburn and Wood (1995) describe a guessing game played on a road network.

Example 3: Player 2 can hide in one of n cells and player 1 has one chance to find him. If player 1 searches the wrong cell, the detection probability is 0. If player 1 searches cell i and the target is indeed there, the detection probability is the given number p_i. For $n=3$ and $\mathbf{p}=(0.25, 0.5, 1)$, the resulting payoff matrix is shown below.

The solution of this game is $\mathbf{x} = \mathbf{y} = (4/7, 2/7, 1/7)$ and $v = 1/7$. In general, the solution of such games can be derived from the principle that x_i and y_i should be inversely proportional to p_i. The $\mathbf{x}$ part of the solution is in some ways surprising. It might be argued that player 1 should favor cells where p_i is large, since he is best at searching such cells. Exactly the opposite is true; a better argument is that player 1 should favor cells where p_i is small because that is where player 2 will probably hide.

	1	2	3
1	0.25	0	0
2	0	0.50	0
3	0	0	1.00

Blotto games. An elementary example is given in Section 6.2.2. In general, a Blotto game involves a competition in n areas and a total resource for each of the two players, x for player 1 and y for player 2. Each player secretly partitions his resource over the areas, with (x_i, y_i) being the allocations of players 1 and 2, respectively, to area i. The payoff to player 1 in area i is some given function $A_i(x_i, y_i)$ of the two allocations, and the total payoff is just the sum of the individual payoffs. In most applications the payoff at target i is the probability that the ith target is killed, with the allocations of the two players being attackers and defenders, respectively. The meaning of the sum is then "average number of targets killed." The rapid proliferation of strategies with n, x, and y makes a direct approach unattractive, but efficient solution methods are nonetheless known. The general idea is to construct n small solutions, one for each target, rather than a single large solution (Beale and Heselden, 1962 , Washburn, 2003b, Chapter 6).

Most interest in Blotto games comes from the Cold War problem of defending a group of targets from attack by ICBMs, in which case the tactic of secretly assigning anti-ballistic missile (ABM) defense levels before anything is observed about the attack is known as a "pre-allocated defense." The defense might hope for better, since the defense has the advantage of acting later in time. If the entire vector $\mathbf{x}$ could be observed before dividing up y, the defense could do things like abandoning heavily attacked targets in order to save targets less heavily attacked. But there are practical difficulties with this idea. Although ICBMs can be tracked easily enough, the real threats are the re-entry vehicles (RVs) that they carry. RVs can be maneuverable and are released only a few minutes from impact, perhaps with decoys that will disguise the RVs until very late in the game. Besides, if all ICBMs are not launched at the same time, the known part of the attack may not total x, so there will be the question of how many ABMs to save for the future. Considerations such as these lead to a pre-allocated defense where a certain number of ABMs are set aside to defend target i, regardless of what is observed about the attack or when the attack occurs, as in a Blotto game. In order to simplify the last-minute control problem, the defense essentially abandons whatever advantage might accrue from acting after the offense. A pre-allocated defense still keeps its allocations secret, thus forcing the attacker to guess at how big an attack is actually necessary for any particular target. This makes it more effective than the Prim–Read defense (Section 4.2.4), which also abandons the advantages of secrecy. Eckler and Burr (1972) include extensive coverage of such missile assignment problems.

The important thing here is that large Blotto games, whether motivated by missile allocation or not, are solvable as a practical matter.

Markov games. A Markov game is actually a class of subgames where a pair of strategy choices, in addition to causing an immediate payoff to player 1, leads to a new subgame within the class. Tree games are Markov games, but the Markov class is more general in that the subgames need not be games of perfect information, and might repeat. American football is a good example: each play (each subgame) leads to another play until the game is finally over, with some plays leading to scores and others not. Although each subgame is small enough to solve as a matrix game, the game as a whole is much too large to permit solution by that method. The computational idea is to work with the subgames, rather than the game as a whole, substituting game values whenever a subgame is called for. The alert reader will note that there is a potential difficulty in deciding where to start, since the value of the subgame one is trying to solve might depend on the values of other subgames that have not been solved yet. Indeed, it is easy to exhibit Markov games that have fatal circularities of this sort, but there is usually no theoretical difficulty as long as some condition (the clock, in American football) forces the game to eventually end.

The main military application of Markov games has been to combat between air forces. Modern fighter bombers are sufficiently flexible that they can be assigned many roles, but they can only do one thing at a time, and must take off with weapons appropriate to the mission. There were some very tense naval moments in World War II when fleet commanders had to decide whether to launch defensive fighter aircraft or offensive bombers, with the ultimate payoff depending very much on similar decisions made by the enemy. When two air forces are engaged in extended combat, one can easily imagine subgames on any given day where the two sides try to guess what the other is doing, with a bad guess leading to an unfavorable subgame on the following day. Berkovitz and Dresher (1959) analyze a situation where each side on each day assigns a single type of aircraft to one of three roles: ground support, counter-air, or air defense. The US Air Force's TAC CONTENDER model is more ambitious, including more missions and aircraft of multiple types. Bracken et al. (1975) review solution methods and propose some approximation procedures.

Continuous Games. Here we assume that the payoff is a continuous function $A(x,y)$, of the two players' strategies. Under suitable conditions (Washburn, 2003b) optimal mixed strategies still exist, as well as a game value. Such games clearly cannot be represented as matrix games, but solutions are nonetheless surprisingly common.

Example 4. Suppose that x and y are each confined to the unit interval, and that player 1 wins if and only if $|x-y| \le 0.2$. The value of this game is 1/3. Player 2 can enforce this value by choosing y in the set {0, 0.5, 1} equally likely. There is no choice of x for player 1 that covers two of these points, so the game's value cannot exceed 1/3. An optimal strategy for player 1 is to choose x in the set {0.2, 0.5, 0.8} equally likely. Three subintervals of length 0.4, each one centered on one of those points, completely cover the unit interval. Any point y in the unit interval must therefore be within 0.2 of one of those numbers, so there is no way for player 2 to avoid all three of them in choosing y, even if he knows the set. Therefore player 1 can guarantee that the payoff will be at least 1/3, so the value of the game must be 1/3. Note that the value is not 0.4, the fraction of the unit interval that player 1 can cover. This is the influence of end effects. Note also that neither player utilizes a continuum of choices, even though such choices are feasible, and that the optimal strategies are not unique.

The "method" used in solving Example 4 was to simply announce the solution and then demonstrate its optimality. Ruckle (1983) defines a class of games called "geometric games," and solves many of them by this technique. There are also continuous games where the players utilize a full continuum of choices, notably duels where each player must choose when to begin firing. There is little to emulate here, since the secret is to be clever enough to see the solution in the first place, but there is one subclass of continuous games where a more systematic approach is possible.

If $A(x,y)$ is a concave function of x for each y, then it can be shown that any mixed strategy for player 1 is dominated by the pure strategy of always using the mean of the mixed distribution, that is, player 1 has no use for mixed strategies. It follows that $v = \max_x \min_y A(x, y)$, where the maximum and minimum are taken over whatever sets are feasible for x and y. Similarly, if $A(x,y)$ is a convex function of y for each x, then player 2 has no use for mixed strategies and $v = \min_y \max_x A(x, y)$. If both conditions hold, we have a concave–convex game where the $\max_x \min_y$ and $\min_y \max_x$ values are equal, i.e., the game has a saddle point. The same arguments turn out to be valid in multiple dimensions.

Suppose there are n opportunities or areas for player 1 to take action, one of which he must choose, and that player 2 must defend against all of them. If player 1 chooses area i and player 2 has defended that area to the level y_i, then the payoff is some given convex function $f_i(y_i)$. Player 2 must defend against all possibilities, but he has only a limited total budget y, which much be partitioned into the n areas. The areas might be drug smuggling routes, opportunities for terrorist attack, biological agents, regions of the ocean in which a submarine might patrol, or independent routes over which a shipment might be made. This class of games is named after the latter –" logistics games." If x_i is the probability that player 1 chooses area i, then the average payoff is $A(\mathbf{x},\mathbf{y}) \equiv \sum_i x_i f_i(y_i)$. Since this is a convex function of $\mathbf{y}$ for all $\mathbf{x}$, the value of the game is the same as the $\min_y \max_x$ value – player 2 selects $\mathbf{y}$ as if player 1 knows $\mathbf{y}$ when selecting $\mathbf{x}$, even though player 1 does not really have that privilege. Since selecting $\mathbf{x}$ when $\mathbf{y}$ is known amounts to choosing the area with the largest payoff, this is essentially worst-case analysis from player 2's standpoint. The optimal $\mathbf{y}$ is the solution to a minimization problem with variables **y** and v, the game value. Specifically, player 2 should minimize v, subject to the constraints that **y** is feasible and $f_i(y_i) \le v; i = 1,...,n$.

An optimal **x** will not emerge from player 2's minimization, but an optimal **x** nonetheless exists. That is, there is some probability distribution **x** such that, even if player 2 knows **x**, he cannot find any **y** such that $A(\mathbf{x},\mathbf{y}) < v$. Player 1 must randomly select an area from this distribution and keep the selection secret from player 2; if he does not keep it secret, then player 2 might discover the selected area and allocate all of his resources to it, thus making the payoff smaller than v.

Example 5. Player 1 will choose one of four areas in which to attack player 2. There will be no damage if the attack is interdicted, but otherwise the damage from an attack in area i is d_i, where $\mathbf{d} = (1,2,3,4)$. The noninterdiction probability is assumed to be $\exp(-y_i)$ (this is the random search formula – see Section 7.3.2), so the average damage to player 2 if player 1 selects area i is $f_i(y_i) = d_i \exp(-y_i)$. If player 2 has a total resource $y = 3$ to distribute to the four areas, how should he do so? The solution is on sheet "Logistic" of *Chapter6.xls*, in the form of a minimization problem (Tools/Solver…) that is solved after the inputs are adjusted. The value of the game is 1.06, and the optimal **y** is (0, 0.63, 1.04, 1.33). Most of the defense goes to area 4 where the prospects of damage are greatest. Area 1 is not defended at all (nor should it be attacked, since the potential damage there is smaller than the game value). If you wish, adjust y and re-solve the game to see how different defense budgets are allocated.

Another example of a logistics game is in Section 8.5.3, where there is an application to a kind of mine warfare.

Logistic games are the simplest, but not the only application of concave–convex games to military affairs. As long as player 1 simply chooses an area, considerable complication is possible in player 2's allocation problem. Player 2 may assign his forces to "missions," rather than directly to areas, with each mission achieving partial effects in multiple areas. Aircraft assigned to search in one area, for example, may have no alternative but to overfly other areas while they are in transit. Player 2 can also have multiple kinds of forces, with a separate constraint for each. Similar games can also be posed on networks, where one player chooses an arc to interdict and the other chooses a path through the network, hoping to avoid the interdicted arc. As in logistic games, the solution can in all cases be found through optimization (Washburn, 2003b).

6.2.4 Information and Its Effects on Combat

One of the most important and difficult problems in modern military analysis is to realistically incorporate the effects of information and uncertainty. Game theory is by no means a panacea, but it does at least deal with the subject. In this section we discuss the problem in general, as well as the possible uses of game theory in dealing with it. We will also conclude the discussion of missile allocation that was begun in Chapter 4.

Consider the finite matrix game (a_{ij}), and let $val(a_{ij})$ be its value. We can also define $\max_i \min_j(a_{ij})$, the best that player 1 can get if player 2 knows player 1's row choice when player 2 selects his column, and $\min_j \max_i(a_{ij})$, the best that player 1 can get if he moves after player 2. These three values can differ substantially, even though they all share the same payoff matrix. They will always be in the order

$$\min_j \max_i(a_{ij}) \le val(a_{ij}) \le \max_i \min_j(a_{ij}). \tag{6.1}$$

For example, the three values for rock-paper-scissors are –1, 0, and +1, respecttively, spanning the range from player 1 losing all the time, to winning as often as he loses, to winning all the time. These strong differences are entirely due to the different information states of the players when they make their decisions.

The strong dependence of the outcome on information is not confined to rock-paper-scissors. Section 9.2.2 includes an analysis of a situation where a searcher (player 2) attempts to detect an infiltrator (who is player 1, since his survival probability is the objective function) using an unmanned aerial vehicle (UAV). The UAV must be within the range of ground stations in order to communicate with the controller. We can identify three decisions that need to be made, two for the searcher and one for the infiltrator:

1. *Z*, the locations of the searcher's ground stations
2. *X*, the search plan for the UAVs
3. *R*, the infiltration route

Let $P(Z,X,R)$ be the infiltrator's survival probability – this function is specified in Section 9.2.2, but its precise nature is not important for the moment. Whether the infiltrator actually survives depends on who knows what when, as well as this function. We can identify several possibilities:

1. A TPZS game where (*Z*,*X*) for the searcher and *R* for the infiltrator are chosen in mutual ignorance.
2. A TPZS game where (*Z*,*X*) is chosen without knowing *R*, and then *R* is chosen knowing *Z* (the ground stations might be visible), but not knowing *X*.
3. A TPZS game where (*Z*,*X*) is chosen without knowing *R*, and then *R* is chosen knowing (*Z*,*X*). This amounts to worst-case analysis for player 2, the searcher.
4. A TPZS game where *Z* is chosen without knowing *R*, then *R* is chosen knowing *Z*, but not *X*, then a spy might discover *R* and report it to player 2 (the spy succeeds half the time, say), and finally

player 2 chooses X, possibly knowing R and possibly not, depending on the spy report.

5. A TPZS game where Z is chosen without knowing R, then R is chosen knowing Z, but not X, and finally X is chosen knowing R and (of course) Z.

There are many other possibilities, some of which incorporate decoys, feints, and erroneous measurements. All of these situations are perfectly well-defined finite TPZS games, so a solution exists in all cases, but most of them are too complicated to analyze. To solve game number 1 as a matrix game, we would have to begin by listing all of the possibilities for (Z, X), itself a large set, and then calculate the survival probability for every entry in the payoff matrix. Game 5 is a tree game, but a discouragingly large one. Game 4 might be the most realistic of those listed, but even describing a strategy for player 2 in that game is a challenge, much less listing them all. Of the five games, only the third is easily solved, and that because the max–min problem can be cast as a simple optimization, as shown in Section 9.2.2.

The example in Section 9.2.2 is sufficiently small that some of the numbered games besides 3 *could* be solved with sufficient effort, but nonetheless two points should be understood. First, when information is involved, it is easy to formulate realistic problems that are simply too difficult to solve, in spite of being perfectly well defined. As a result, we often end up deliberately solving the "wrong" problem as a way to at least gain some insight into the "right" one. The solution of game 3, for example, is at least a lower bound on the value of game 1. Second, the value of a game can be distressingly sensitive to who knows what when. That is, we should not lose sight of the fact that the wrong problem that we can actually solve has significant differences from the right one. When it comes to dealing with information, humility is a good stance for a combat modeler. The next case study makes the same point in a different, more important problem.

JOINT DEFENDER, A Case Study. The USA has some ICBM interception capabilities, some of which are anti-ballistic missiles (ABMs) located on ships. These assets might be of use in a scenario where a rogue state has chosen to launch an ICBM attack on the USA or one of its allies. If there is sufficient political warning of such a possibility, we might consider the question of where such ships should be located in order to optimize their capability for frustrating such an attack. In addition to providing a warning, the political situation might also identify the potential ICBM targets, and even perhaps the value of each target on some relevant scale, so we can say that the objective is to deploy the ships so that the average total value killed by the hypothetical ICBM attack is minimized. The collection of ship locations is called $\mathbf{X}$.

The rogue state (player 1) has some choices to make that also influence the outcome. The rogue state is supposed to be sufficiently large in extent, compared to the range of its ICBMs, that the location of the ICBMs within its territory is significant. Before the attack, the ICBMs may be transported from one place to another. The transportation plan is called $\mathbf{W}$. It will also be necessary, of course, to decide which ICBM should be aimed at which target. The aiming plan is called **Y**.

Although our main purpose in modeling is optimal ship deployment, the actual assignment of ABMs to the ICBMs is also relevant to the chosen effectiveness measure. The ABM assignment plan is called **R**. Let $F(\mathbf{X},\mathbf{R},\mathbf{W},\mathbf{Y})$ be the average value killed, given all of the decisions represented by the four arguments.

Chronologically, the decisions happen in the order **W**, **X**, **Y**, and finally **R**. We might therefore expect the object of computation to be $\max_{\mathbf{W}} \min_{\mathbf{X}} \max_{\mathbf{Y}} \min_{\mathbf{R}} F(\mathbf{X},\mathbf{R},\mathbf{W},\mathbf{Y})$, that is, **W** is selected knowing nothing about **X**, **Y**, or **R**, then **X** is selected knowing only **W**, etc. This situation might be imagined as a game of perfect information, a very large tree game where the root node is a max that branches on **W**, then each branch goes to a min node that branches on **X**, and so on. But it does not follow from chronology that the minimizer knows **Y** when selecting **R**, since ABMs will have to be committed with only partial information about which targets are being attacked, and there is good reason for the maximizer to keep **Y** secret. The actual object of computation will have to depend on exactly what is assumed about who knows what when.

Brown et al. (2005) describe JOINT DEFENDER (JD), a model dealing with the situation described above. JD is a state-of-the- art minimization that utilizes 2005-era computers to their capacity, so the reader might expect that JD will make realistic assumptions about the information state of each participant when decisions are made. That is not the case, as can be seen by comparing reality with JD's assumptions. In reality, movements of both ships and ICBMs are subject to disguise, so we might suppose that **X** and **W** are selected in mutual ignorance, at least partially. JD instead gives the advantage to the rogue state, assuming that **X**, along with all ship characteristics such as the number of ABMs aboard, is known when **W** and **Y** are selected.

In reality, **Y** is to some extent known when **R** is selected, since there will be radar and satellite observations of ICBM tracks when ABMs must be committed. In a Blotto model of this, we might assume that ABMs are randomly but secretly committed in advance to defend certain targets, ignoring all tracking information except the part that is required to guide an ABM to its target. JD is even more pessimistic, assuming that the ABM preallocation cannot be kept secret, that is, **W** and **Y** are selected knowing **R**. Thus, instead of **R** being partially informed about **Y**, in JD **Y** is perfectly informed about **R**. Again, JD gives the advantage entirely to the rogue state.

JD's final favor to the rogue state is its assumption that each ABM must be committed to defending against an attack by a particular missile on a particular target, and will not be launched unless that attack occurs. To emphasize the extreme pessimism of this assumption, suppose the attacker has one perfect ICBM and 10 targets to choose from, while the defender has nine perfect ABMs. JD's defender cannot adopt the policy of shooting down the lone ICBM regardless of what it attacks, but must instead preallocate each ABM to defending one of the targets, and then announce the commitments to the attacker. As a result, the attacker will always find an undefended target and kill it.

What are the analytic advantages of JD's pessimistic (for the minimizer) assumptions? The value killed when both sides optimize will be $\min_{\mathbf{X}} \min_{\mathbf{R}} \max_{\mathbf{W}} \max_{\mathbf{Y}} F(\mathbf{X},\mathbf{R},\mathbf{W},\mathbf{Y})$, which can be written as $\min_{\mathbf{X},\mathbf{R}} \max_{\mathbf{W},\mathbf{Y}} F(\mathbf{X},\mathbf{R},\mathbf{W},\mathbf{Y})$ to emphasize that this is basically worst-case analysis – the minimizer selects **X** and **R** expecting the worst possible attack for the selected values. The sets of alternatives are much larger than in the UAV problem discussed above, but JD employs the same worst-case criterion. In JD, the sets of alternatives are so large that even the min–max problem is not directly solvable, so it is necessary to approximate $F(\mathbf{X},\mathbf{R},\mathbf{W},\mathbf{Y})$ with a linear function, after which the inner maximization can be replaced by an equivalent minimization based on the dual of a linear program. The problem is thus reduced to finding the minimum of something over a large set of alternatives (see Brown et al. (2005) for details), a very large but at least conventional problem that JD can solve exactly.

The JD solution will include the optimal **X**, which is the best way of stationing the ships. It will also include the optimal **R**. It would not be reasonable to advocate actually using **R** in the real world – why artificially restrict the defender to the type of policies considered? However, it might be reasonable to advocate using **X**, especially if the defender is so preponderant that JD can guarantee a satisfactory result, in spite of all its pessimism. JD is an advisor in a world where advisors are scarce, so its "optimal" solution is at least worth studying.

In an ideal analytical world, we would incorporate the correct expression of the uncertainty that surrounds decisions into all our models and optimize or solve games as appropriate. That is usually not possible, so we are forced to employ approximations and artificialities of the type used in JOINT DEFENDER. We "see through a glass darkly," but it still pays to keep looking.

6.3 Wargames

A wargame is a more or less abstract model of warfare that permits humans to play the role of combatants. There are many forms. Commercial wargames often take the form of a playing board representing a part of the Earth on which pieces representing military units are moved around. Military wargames (and to an increasing extent commercial wargames as well) often rely on computers to enforce realism in unit movements, to assess damage, and to display whatever parts of the game's status that the rules permit each player to see. Some wargames incorporate nonplaying humans (referees) to help control the game's evolution, and some do not. Some are very simple, while others are complex, detailed, and can require several days to play.

Most wargames are played for recreation, and there is a significant community of hobby wargamers. Several companies publish wargames commercially, and there are even magazines (*Fire and Movement*, *Strategy and Tactics*) devoted to the activity. There has been considerable interplay between hobby and professional wargamers, even though the audiences are distinct. Perla (1990) describes some of the innovations that both groups have introduced.

Professional wargaming (wargaming carried out by or for the military) has two major purposes. The paramount purpose is education and training. Wargame participants must make decisions under stress with incomplete or even false information, and must cope with the capricious effects of luck and enemy action. These characteristics of warfare are difficult to simulate in any way other than in a dynamic, competitive simulation that involves humans. The experience of participating in a realistic wargame is often very intense, so that lessons learned are long remembered. At the end of WWII, Adm. Nimitz stated that

> *The war with Japan had been re-enacted in the game rooms [at the Naval War College] by so many people and in so many different ways that nothing that happened during the war was a surprise - absolutely nothing except the Kamikaze tactics towards the end of the war; we had not visualized these.*

All three US services, and most other military organizations, utilize wargaming as an educational device.

The teaching power of wargaming has its dark side. If the structure of the game being played reflects politically correct assumptions that are not realistic, then the lessons learned may be useless or even worse. Bracken (1976) states that the initial invasion of Russia in 1941 was so successful because Stalin had decreed that the only tactic that could be included in wargames (both sides wargamed the invasion beforehand) was a defense of the border. The Soviet Union eventually perfected the defense-in-depth tactics that were actually needed, but the early cost in terms of captured divisions was terrible. Also, wargames will inevitably ignore certain aspects of reality, and to some extent will thereby teach that those aspects are not present in the real world. Fear and panic are not experienced in wargames, for example. Most wargames in the United States are carried out using communications in English, where nobody runs out of water or toilet paper. That may not actually be the case.

Wargaming can be used for analysis as well as education and training. Given the capability to fight abstract wars, it is tempting to test the effect of different weapons or tactics on the wargaming outcome, perhaps basing procurement decisions on results. The tendency to use wargaming in this manner has fluctuated in the past. Wargaming's utility as an analytical tool is not as obvious as in its use in education, so the technique will probably continue to come in and out of favor.

The major recommendation for using wargaming as an analytic tool is that it gets around most of the difficulties mentioned earlier in discussing game theory. Game theory requires that payoffs be explicit functions of the strategies. In wargaming, on the other hand, one can argue that the payoffs come with the backgrounds of the players, so that a brief statement about goals ("Take Stalingrad, but whatever you do don't Third Panzer Division") will suffice. In game theory, the need to initially list all possible ways in which a battle might evolve is often so daunting as to either inhibit the exercise or lead to oversimplification.

For this and other reasons, abstract games are often unsolvable, while questions of solvability do not arise in wargaming because "optimal" tactics are not sought. Since wars are fought by humans, it makes sense to use human players in emulating combat, whatever their weaknesses and motivations. If the results reflect those weaknesses and motivations, then that, the argument goes, is as it should be.

The counterarguments are of two kinds, the first having to do with motivation. A wargame is only a more or less accurate abstraction of the world. Players sometimes express a felt conflict between making decisions as they would be made in the real world and making decisions in order to do well in the abstraction. These decisions may differ on account of readily perceived modeling deficiencies. For example, the scoring of a wargame may not be affected by casualties incurred, but only by success in achieving an objective, whereas casualties are always of concern in reality. How should those players be advised? The answer is not obvious, and the resulting murkiness does not augur well for making decisions about tactics or equipment based on the results of play. Player attitudes toward risk are particularly worrisome; there is a difference between having a blip labeled "CVN" disappear from a computer screen and having an actual aircraft carrier sunk. In games where there are possibilities for coalition formation or cooperation, it is questionable whether the feelings about honor and guilt that would accompany certain actions in the real world can be reproduced in an artificial situation.

The second counterargument raises technical objections about using wargaming as a research technique. The idea that it is best to capture decision making that is typical, rather than optimal, is particularly questionable in research games that represent an unfamiliar reality to the players. There is a natural tendency to overstate the effectiveness of new weapons or tactics when played against an inexpert enemy who has yet to develop countermeasures. The enemy might very well develop some tactical countermeasures in the process of repeatedly playing the wargame, but wargaming is sufficiently time consuming and expensive that repeated play of the same game by the same players is rare. To some extent this lack of repetition is deliberate, since research wargamers often try to avoid player learning because they fear that the learning will be about the wargame's artificialities. As a result, questions of statistical validity are seldom dealt with. Deliberately seeking enemy play that is typical, rather than optimal, violates a well-known military maxim by designing to the enemy's intentions instead of the enemy's capabilities. Finally, typical human play often differs substantially from optimal play in TPZS games for which the solution is known (Kaufman and Lamb, 1967). Thomas and Deemer (1957) downplay the usefulness of wargaming as a tool for solving TPZS games, although they admit its usefulness for getting the "feel" of a competitive situation. Berkovitz and Dresher (1959) state bluntly that "In our opinion ... [wargaming] is not a helpful device for solving a game or getting significant information about the solution."

The counter-counterargument, of course, is that, whatever the objections to it, research wargaming is often the only feasible technique for examining a genuinely competitive situation. Theoretical and experimental approaches each provide insights into combat, but a great deal of uncertainty must remain.

Exercises

(1) In Section 6.2.1, the distribution $\mathbf{y} = (4/19.\ 15/19)$ is claimed to have the property that both of player 1's strategies lead to a kill probability of 7/38. Show that this is the case.

(2) In Section 6.2.1, it is announced that player 1's optimal strategy $\mathbf{x} = (3/19,16/19)$ can be derived by equalizing the average payoff for both of player 2's strategies. Show that this is true by equating the payoff for both SH and DP to the unknown game value v. Solve the resulting equations for $\mathbf{x}$ and v.

(3) Rock-paper-scissors is a TPZS game with a circular domination rule: Rock beats scissors beats paper beats rock. Model it as a 3×3 game where the payoff is the net number of times that player 1 wins (1 if his choice dominates, –1 if his choice is dominated, otherwise 0). Show that the value of this game is 0, and that the optimal mixed strategy for both sides is (1/3, 1/3, 1/3).

(4) Reconsider Example 2. Give Blotto four regiments, instead of three, and solve the resulting game.
Ans. Blotto and his enemy are each equally likely to choose any strategy, and the game value is 0.8.

(5) Solve the following 4×5 example of a guessing game. The strategies might be cells in which player 2 might hide, or cells in which player 1 might search

	1	**2**	**3**	**4**	**5**
1	3	1	0	0	0
2	1	2	2	0	0
3	0	0	3	0	1
4	0	0	0	2	3

Ans. $\mathbf{x} = (0.17,\ 0.34,\ 0.06,\ 0.42)$, $\mathbf{y} = (0.29,\ 0,\ 0.29,\ 0.43,\ 0)$, and $v = 0.857$.

(6) Consider the logistics game where $n = 3$, and $f_1(y_1) = 3/(1+y_1)$, $f_2(y_2) = 2/(1+y_2)$, and $f_3(y_3) = 1/(1+5y_3)$. Using sheet "Logistics" of *Chapter6.xls* as a model, construct a spreadsheet that uses Solver to find the best way for player 2 to partition a budget of y over the three areas, with y being an input. If $y = 2$, what is the best $\mathbf{y}$, and what is the value of the game?
Ans. $\mathbf{y} = (1.4, 0.6, 0)$, and $v = 1.25$.

(7) Consider a Blotto game where there are six attackers, eight defenders, and four targets. All attackers and defenders are perfect, so each target will be killed if and only if the attackers outnumber the defenders. The payoff is the average number of targets killed.

(a) What are the max–min and min–max values?

(b) Consider the following tactic for the defender. Defend the targets in pairs, with each pair getting four defenders. Within each pair, assign a number of defenders to one target that is equally likely to be any number between 0 and 4, and the rest of the defenders to the other target. This tactic guarantees that the average number of targets killed will not exceed … what?

Ans. The max–min value is 0, and the min–max value is 2 (achieved by defending each target with two defenders). In part b, the suggested randomized tactic guarantees that not more than 1.2 targets will be killed on the average. That strategy turns out to be optimal for the defender, that is, the value of the game is 1.2.

Chapter 7
Search

If Edison had a needle to find in a haystack, he would proceed at once with the diligence of the bee to examine straw after straw until he found the object of his search... I was a sorry witness of such doings, knowing that a little theory and calculation would have saved him ninety per cent of his labor.

Nikola Tesla (1931)

7.1 Introduction

First of all, the subject of this chapter is not searching on the internet. Nor is it searching for a job or inner peace, and it has nothing to do with data mining. These initial caveats are necessary because practically all intellectual activity can be thought of as a search for something, so the term "search" by itself implies very little. It is an outright mistake to say "search" to a search engine, and even "search theory" will expose many entries having nothing to do with what we have in mind.

This chapter is about searching for physical objects with sensors that act very much like our eyes and ears. Eyes sense electromagnetic radiation, as do radars, lasers, and cameras. Ears sense sound vibrations, as do microphones and sonars. The fundamental phenomena in each case are such that signal strength decreases rapidly with distance from the source, so physical proximity is the primary requirement for detection. Details about specific sensors can be found in books such as Skolnik (2001) for radar or Urick (1996) for sonar, but will not be needed here. The only important point for our purposes is that proximity is required for detection.

It is true that we can detect stars over astronomical distances, but that is only because they are very strong sources exposed against a background with very little noise. Targets on Earth, especially those of military interest, are usually not so cooperative, so searching often involves motion as the searcher tries to achieve the required proximity. The searcher will frequently move from place to place or hope that his target does so. Roughly speaking, search is a sequence of repeated trials where successes happen when the distance between searcher and target is short enough.

Part of the reason for the increasing importance of search is that targets have an increasing tendency to be hidden. During the Vietnam War, the US Air Force spent 800 aircraft sorties and lost 10 aircraft in futilely trying to destroy the Thanh Hoa Bridge near Hanoi. After the introduction of laser-guided bombs, the same

A. Washburn, M. Kress, *Combat Modeling*, International Series in Operations Research & Management Science 134, DOI 10.1007/978-1-4419-0790-5_7,

bridge was destroyed with only four additional sorties (USAF, 2005). The increasing lethality of precision-guided weapons has made it unsafe to be a well-located, obviously hostile target, so it should not be surprising that targets of military interest are increasingly hard to locate and identify.

Some forms of warfare have always been nearly pure search problems. The effectiveness of a minefield depends very much on the locations of the constituent mines being unknown, so clearing a minefield is essentially a search problem. The whole point of being a submarine is to (nearly) prevent the successful operation of electromagnetic sensors, thus forcing a difficult search problem on antisubmarine forces.

The great questions in search all involve time. We ask, "How long will detection take?" or "What is the probability of detection in a fixed time?" Detection is inevitable, given sufficient time. The object of search planning is to speed things up.

Search effort may either be applied continuously, as when a man walks around on the sidewalk looking for his car keys, or discretely, as when an aircraft drops a sequence of sonobuoys, each of which examines a certain section of ocean. In the discrete case, search theory has much in common with the firing theory of Chapter 2. In both cases a sequence of discrete attempts (shots in Chapter 2, looks in this chapter) is made to achieve a success (kill in Chapter 2, detection in this chapter), so the mathematics of Chapter 2 can sometimes be applied to search problems.

Example 1: A submarine has been located within a location error that has a circular normal distribution with standard deviation 500 m. An aircraft has four reliable sonobuoys available, each of which has a detection radius of 200 m, and each of which can be delivered to any aim point, except that each sonobuoy delivery has a circular normal delivery error with standard deviation 100 m. In what pattern should the four be aimed, and what is the resulting detection probability? To approximately answer this question, we might employ the diffuse Gaussian analysis of Section 2.3.3, with its accompanying sheet "DGPattn" of *Chapter2.xls*. The area covered by each sonobuoy is $\pi(200)^2$. Since the lethal area of a diffuse Gaussian weapon is $2\pi b_X b_Y$, we equate the two areas to find that $b_{X_i} = b_{Y_i} = 141.4$ m for $i = 1, 2, 3, 4$. The necessity of equating lethal areas is what makes this analysis an approximation. Since the delivery errors are circular, take $\sigma_{X_i} = \sigma_{Y_i} = 100\,\text{m}$ for $i = 1, 2, 3, 4$. Since the submarine's location error is circular and common to all shots, take $\sigma_U = \sigma_V = 500$ m . If this data are entered onto sheet "DGPattn", using a reliability of 1 for each shot because the sonobuoys are all assumed to function as designed, and if the aim points are then optimized using Solver, the result is that the four sonobuoys should be aimed in a square of side approximately 416 m, and the associated detection probability is 0.22. This is referred to as "P_K" in Chapter 2, but in the current application it is a detection probability.

When search is performed continuously, rather than discretely, we need a different notion of equivalency than the lethal area of Chapter 2. This is the notion of sweep width, developed in the next section.

7.2 Sweep Width

The simplest description of a sensor is to give its detection range R, the distance within which detection will surely happen, or else surely not happen if the distance is larger than R. We might say of the unaided eye that its detection range for a tank in open terrain in daylight is one mile. Unfortunately, while a fixed detection range satisfies our desire for a simple, proximity-based measure of a sensor's performance, it is frequently too crisp an idea. The eye's detection range might depend on the tank's speed and camouflage, the turret orientation, the platform the eye is mounted on, the presence of smoke or haze on the battlefield, and other factors. The idea of a fixed or "cookie-cutter" detection range is so appealing that we will continue to use it, but attempts to forecast fixed ranges in the real world are often disappointing. Forecast detection ranges for sonars are notoriously subject to error – it is not uncommon to be off by a factor of two or more.

In Chapter 2, a similar indeterminacy problem was solved by introducing the damage function $D(r)$, whose meaning was the probability of kill if the weapon detonates at range r from the target. We then defined a notion of equivalency based on lethal area, as employed in Example 1 above. That tactic will not work for continuously moving sensors. The problem is that the notion of a "trial" is not well defined. If we let $D(r)$ be the detection probability "at" range r, we will then have to deal with an infinity of closely spaced trials, since the sensor looks continuously. Detection will be nearly certain if all those trials are assumed to be independent. Rather than try to deal with such difficulties, we instead proceed by defining a basic search experiment that may or may not result in detection. This is the approach originally used in World War II by the Operations Evaluation Group (OEG, 1946).

The basic experiment is as illustrated in Figure 1. Relative motion between sensor and target is assumed to be at constant velocity in two dimensions, characterized by a lateral range.

Definition: If relative motion between sensor and target is assumed to be at constant, nonzero velocity in two dimensions over a complete line, the *lateral range* is the distance between searcher and target at the point where the distance is minimal, taken to be positive if the target lies to the searcher's left at the minimal point or otherwise negative. The minimal point is also called the *point of closest approach.*

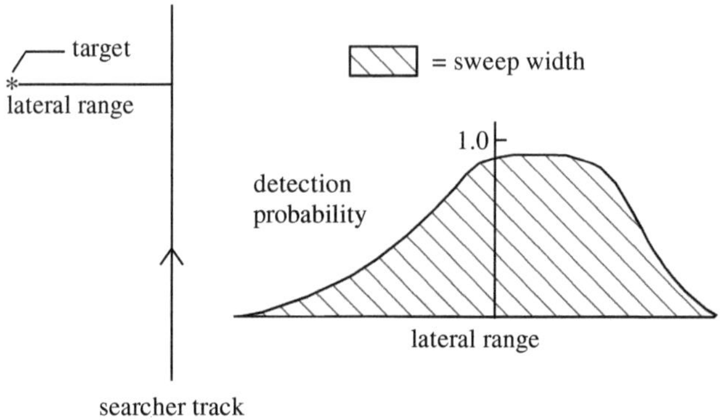

Figure 1: Illustrating a top view of the basic experiment on the *left*, shown with a positive lateral range, and an asymmetric lateral range curve on the *right*.

If the experiment of moving over a complete line is carried out, it may or may not result in a detection. We do not attempt to keep track of the time or "true range" where the detection occurs; it may even occur after the point of closest approach. This might happen if, for example, the searcher's eyes no longer have to cope with glare from the sun once he is past the target. Any detection at any point on the line constitutes a success for the experiment. Let $P(r)$ be the detection probability at lateral range r.

Definition: A graph of detection probability $P(r)$ versus lateral range r is a *lateral range curve*. The area underneath the lateral range curve is called the *sweepwidth.*

Note that a sweepwidth has dimensions of length in spite of being an "area," since the vertical dimension of a lateral range curve is dimensionless.

Lateral range curves are usually symmetric about 0, unlike the curve shown in Figure 1, so it is not usually necessary to remember the left/right sign convention for lateral ranges.

Now imagine that search consists of a series of "passes," each of which can be thought of as movement over a whole line with its own lateral range. The lateral ranges themselves will be random, since the location of the target is unknown. Let $f(r)$ be the density function of the lateral range for some pass. By averaging over the lateral range, we obtain the probability of detection P_D for that pass:

$$P_D = \int_{-\infty}^{\infty} f(r)P(r)dr \cong f(0)\int_{-\infty}^{\infty} P(r)dr = f(0)W\,, \tag{7.1}$$

where W is the sweepwidth. The approximation in (7.1) is accurate as long as the density $f(r)$ is nearly constant over the interval where $P(r)$ is significantly positive, in which case the density can be factored out of the integral. This will usually be the case when lateral ranges tend to be much larger than the sweepwidth; that is, when each pass results in a small detection probability. The integral in (7.1) is then the area under the lateral range curve or the sweepwidth. In this sense, the sweepwidth is all we need to know about the sensor. If we have to state a single number that describes the effectiveness of a continuously moving sensor, the sweepwidth W is the most natural one to give.

If we actually had a definite range sensor that would detect targets inside range R, but not otherwise, then the lateral range curve would be 1 if for $|r| \leq R$, or otherwise 0, and the sweep width would be $2R$. We thus have a notion of equivalency, albeit a different one than the notion of Chapter 2. A sensor with sweepwidth W is equivalent to a sensor with definite detection range $W/2$, as long as search is of the type for which (7.1) correctly represents the detection probability per pass. The last caveat makes equivalency a risky assumption in many of the situations where we would like to employ it, but sweepwidth is still a valuable notion.

Sweepwidths can be measured by repeatedly performing the basic experiment. The U.S. Coast Guard has great interest in the detection of marine targets from both air and sea platforms and has performed numerous experiments whose ultimate goal is to estimate sweepwidth (e.g., Robe et al., 1985). Tables of visual sweepwidths for marine targets are available (e.g., NSARC, 2000).

Sometimes it is the target that moves, rather than the sensor, but the same theory applies. The interaction between stationary naval mines and their moving ship targets is usually imagined to be as in the basic experiment, effectively summarized by recording a lateral range curve and its corresponding sweepwidth.

7.3 Three "Laws" for Detection Probability

We now consider the search of a fixed region with area A that is known to contain a single stationary target, with no prior knowledge of the target's location. A single searcher looks continuously for the target, moving at speed V and with fixed detection radius R. The assumption that no prior knowledge is available means that the target's location density is constant (necessarily $1/A$) within the region. We might have a ship looking for a lifeboat, a tank looking for its next target, a satellite looking for a particular truck, or a man trying to find his car keys. The important parameters are A, V, R, and the amount of search time t. Depending on how the search starts, one might argue that a circle of radius R is covered initially and instantaneously, but we will make no allowance for any resulting initial detections in the following. We will refer to the sweepwidth $W = 2R$, rather than R, even though our initial analysis concerns only a fixed detection radius.

After a time t, the searcher will have covered a region of length Vt and width W, the area of which is VWt. The ratio of this area to the area of uncertainty A plays a central role.

Definition: The *coverage ratio* z is the ratio VWt/A, the ratio of the area covered to the area of uncertainty.

We are usually interested in the detection probability as a function of time, which we call $P_D(t)$. In problems where search continues until the target is found, let T be the random time until the first detection. Since $P_D(t) = P(T \le t)$, the mean value of T is given by

$$E(T) = \int_0^\infty P(T > t)dt = \int_0^\infty (1 - P_D(t))dt. \tag{7.2}$$

For nonnegative random variables like T, the first equality of (7.2) can be shown to be true using integration by parts. Since the mean time to detection is the area underneath a graph of the nondetection probability, it will be small when the detection probability is large.

7.3.1 Exhaustive Search

The best that the searcher can hope for is that the area VWt is all located within the region to be searched and does not overlap with itself. In that case, since the target is equally likely to be anywhere in a region of area A, the detection probability is VWt/A, the coverage ratio. More precisely, since the detection probability $P_D(t)$ cannot exceed 1, we have

$$P_D(t) = \min(1, z), \text{where } z = VWt / A. \tag{7.3}$$

Formula (7.3) is actually an upper bound, since the searcher cannot hope for better. If the searcher can navigate accurately and does not need to turn too often while staying inside the region of uncertainty, the formula may be accurate. A searcher attempting to achieve an exhaustive search might follow a path that looks something like the path of a lawn mower; indeed, exhaustive search is sometimes referred to as a "lawn mower search" for that reason. The searcher might also employ a spiral of some kind in attempting to cover all of a circular area without overlap.

The maximum time to detection is $A/(VW)$ in an exhaustive search, and the mean time to detection is easily shown to be half of that (Exercise 1).

7.3.2 Random Search

A different kind of approximation is to assume that the covered area will overlap significantly with itself, with the amount of overlap being whatever happens "at random." More precisely, we assume that the covered area VWt is in effect shredded into n pieces of confetti before being distributed uniformly over the region of interest. Each piece of confetti has a chance of $\frac{VWt}{nA}$ of covering the target. Since each piece of confetti is assumed to be an independent trial, we can use the powering-up idea of Chapter 2 to conclude that the overall detection probability is

$$P_D(t) = 1 - (1 - \frac{VWt}{nA})^n \cong 1 - \exp(-z), \text{ where } z = VWt / A. \tag{7.4}$$

The approximation symbol in (7.4) means that n is irrelevant as long as it is large, with $1 - \exp(-z)$ being the limit as n approaches infinity. It does not matter exactly how many pieces of confetti there are, as long as there are lots of them. The important thing is that the total amount of area is effectively cut up into confetti.

Formula (7.4), the random search formula, is the most widely applied formula in search theory, but it seems to correspond to a rather unlikely idea. Why would anyone want to cut his search area up into confetti, even if he could? The justification for the wide use of (7.4) is not that anyone would deliberately search randomly, and certainly not that there is any physical analog to scattering confetti in a real search, but rather that many of the difficulties placed in the way of real searches have the same randomizing effect. Real searches will have overlap, either because of the necessity to turn or because navigation is imperfect. Targets sometimes move slightly as search proceeds, even if they do not intend to, thereby spoiling the perfection of a lawn mower pattern. An exhaustive search can only be performed with a definite range sensor, whereas most sensors have indefinite fields of coverage that force a certain amount of overlap. The effect of all of this is that the intended exhaustive search frequently gets turned into something that is not far in effect from scattering confetti.

When t is very small, formulas (7.3) and (7.4) agree that the detection probability is approximately $(VW/A)t$. The coefficient (VW/A) can be thought of as a constant detection rate λ in a Poisson process of detections, with (7.4) being the probability that there are no detections over an interval of length t. The reader who is familiar with the wide applicability of Poisson processes may consider that observation to be a better derivation of (7.4) than the confetti analogy. The characterizing assumption of such processes is that events in nonoverlapping intervals of time are independent. This is the real difference between (7.3) (which assumes that past failure ensures future success through a process of elimination) and (7.4) (which assumes that the chance of success in the next interval of time is independent of what has happened in the past).

According to (7.4), the time T to detection in random search is an exponential random variable with parameter λ, and its mean is therefore $A/(VW)$, the reciprocal of λ. Thus $E(T)$ is exactly twice as large as in exhaustive search. The time T lacks memory in the sense that $P(T > t + x \mid T > t)$ is independent of t; that is, it does not matter how long you have already been searching unsuccessfully if you are trying to predict when detection will finally occur (Exercise 2). In a sense, searching at random feels very much like spinning a roulette wheel until "00" finally shows up. The wheel cannot remember the past, and neither can a random search.

Figure 2 shows both formulas (7.3) and (7.4), in addition to a third formula that will be introduced in the next section. All three curves start out with the same slope (namely λ), but otherwise differ significantly. They differ the most when the coverage ratio is 1, exactly the amount of time that a searcher who believes he is making an exhaustive search would commit to the task.

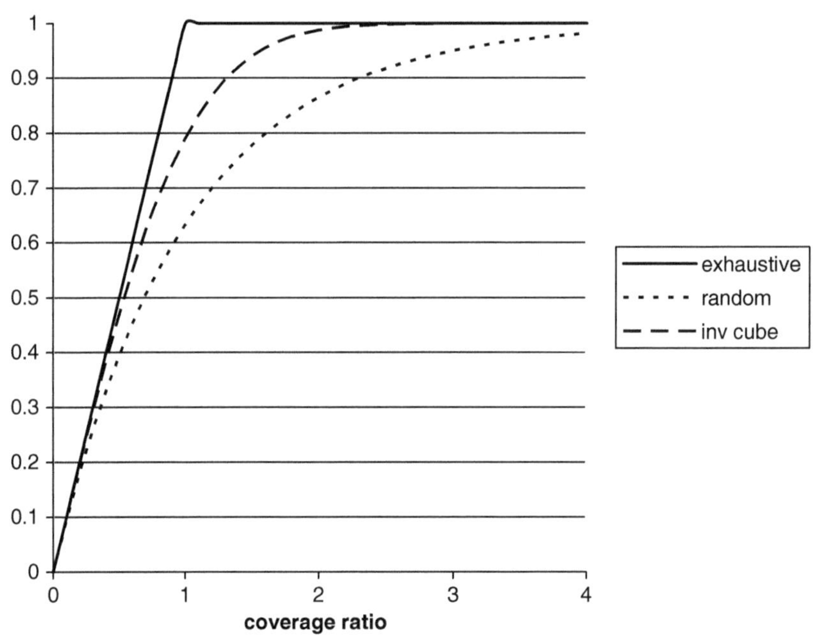

Figure 2: Detection probability versus coverage ratio for exhaustive search, random search, and the inverse cube law.

Even though searching randomly is not usually the goal, but simply a skeptical analytic concession to reality, there are circumstances where a feasible method of actually performing a random search can be useful. These circumstances often involve moving targets (Section 7.6) and may involve robots who need to be programmed to move in some specific manner. The default way of accomplishing this is"diffuse reflection." In diffuse reflection, the path of the searcher has the same statistical properties as the path of a light photon making diffuse reflections from rough walls, hence the name. Consider any convex, two-dimensional region such as a circle or rectangle. Begin anywhere, choose any direction and move at fixed speed in a straight line until the boundary is contacted. After reflecting from the boundary, move at fixed speed in a straight line until the boundary is contacted again, and so on. Each reflection from the boundary should be at a random angle θ measured from the local tangent to the boundary, and the density function of this angle should be $0.5\sin(\theta)$ for $0 \le \theta \le \pi$. Note that this density is *not* uniform in diffuse reflection. Figure 3 shows the kind of searcher track that results.

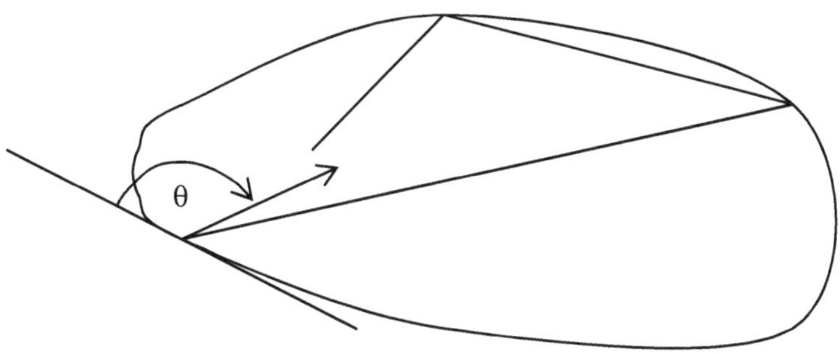

Figure 3: Illustrating diffuse reflection. All movement is at fixed speed and the random reflection angle has density 0.5sin(θ), independent of all other reflection angles.

Lalley and Robbins (1987) show that diffuse reflection has most of the properties that we expect of a uniformly random search. For example, the region is covered uniformly in the sense that, regardless of the shape of the boundary, the amount of time that the searcher spends in any subregion depends only on the area of the subregion, not its shape or location. If for any reason you should want to implement random search, consider diffuse reflection. If you are tempted by something else, then be aware that many kinds of random motion will not cover the region uniformly.

7.3.3 The Inverse Cube Law

The inverse cube law of detection was invented in World War II to deal with situations where a lookout is trying to detect the wake of a ship (the wake of a fast moving ship is much more visible than the ship itself). The solid angle subtended at the eye by such a wake decreases inversely with the cube of the range to the wake as long as the altitude of the eye above the plane of the wake remains constant, hence the name of the law. The essential assumption is that target detection is a Poisson process where the event rate is proportional to that solid angle, with the proportionality constant depending on such things as the target's size, its contrast with the background, and other parameters. The sensor's sweepwidth depends on this proportionality constant.

Consider a lawn mower search of an area where the searcher's parallel passes are separated by a track spacing S. The resulting coverage ratio is W/S, where W is the sweepwidth. The exact formula for the inverse cube law detection probability requires several pages to derive (see OEG, 1946 or Washburn, 2002), so we will not reproduce it here, but the final formula is

$$P_D(t) = 2\Phi(z\sqrt{\frac{\pi}{2}}) - 1, \text{ where } z = W/S. \tag{7.5}$$

Equation (7.5) is the middle curve in Figure 2. Since the track spacing possible when searching a fixed area A at speed V in time t is $S = A/(Vt)$, the definition of z in (7.5) is really the same as in (7.4); i.e., the coverage ratio.

The appearance of the cumulative normal distribution $\Phi()$ in (7.5) is an analytic mystery. No random variable has been assumed to be normal, and yet the cumulative normal function appears! The function is widely tabulated, but cannot be expressed in terms of more elementary functions. When $z = 1$, the detection probability is 0.79, well below the detection probability for exhaustive search, but well above the detection probability (0.63) for random search.

In addition to its derivation as the natural function to use when detecting wakes, the inverse cube law is of interest because it seems to be a good compromise between the optimism of exhaustive search and the pessimism of random search (see Figure 2). Most current interest derives from the latter fact. The International Aeronautical and Maritime Search and Rescue Manual (ICAO, 2003) asks the search planner to decide whether search conditions are "ideal" or "poor" for purposes of calculating detection probabilities, interpreting the former as the inverse cube law and the latter as random search. The manual does not employ the exhaustive formula. Sweep width is obtained from a table, so the planner's choice of the track spacing S determines the coverage ratio and hence the detection probability.

7.4 Barrier Patrol

In this section we assume that the searcher uses a cookie-cutter sensor with definite range R. The definite range assumption is essential, since the searcher will usually follow a highly nonlinear track that contrasts sharply with the basic lateral range experiment of Section 7.2.

Searchers sometimes take advantage of environmental features that force moving targets through narrow restrictions. These restrictions might be straits or harbor mouths at sea or on land they might be mountain passes, watercourses, or (if targets are in wheeled vehicles) roads. By remaining in the vicinity of these restrictions, the searcher hopes to construct a one-dimensional barrier to movement, rather than deal with a large, two-dimensional expanse. Patrolling near Gibraltar, for example, was a tactic used to prevent U-boats from entering the Mediterranean in World War II.

Let L be the width of the barrier to be protected. We assume that target motion is perpendicular to the barrier. Consider a target whose speed is U and assume that the barrier penetration point is uniformly distributed over L. If $2R$ is larger than L, then a single, stationary sensor in the middle of the barrier is sufficient. Otherwise, let the searcher's speed be V. The searcher must remain in the vicinity of the barrier, so his track will consist of repetitive traverses of a closed curve (see Figure 3). What kind of closed curve is best, in the sense of producing the highest detection probability?

The answer to the last question is unknown, but there is good reason to suppose that a particular upper bound is actually a good approximation to the best possible detection probability. To derive it, first imagine that there are actually many targets uniformly distributed over a tape that moves past the searcher at speed U. A point on the tape is "covered" if it ever comes within distance R of the searcher, and the searcher's goal is to cover the largest possible fraction of the tape. This is a simple rephrasing of the question, since the fraction of the tape covered is the same as the probability of detecting any particular target on it.

If θ is the angle between the two velocity vectors, then the searcher's speed relative to the tape is $S = \sqrt{U^2 + V^2 - 2UV\cos(\theta)}$. The relative speed S will vary with time because θ varies with time as the searcher moves around the closed curve. However, since the average value of $\cos(\theta)$ must be 0 because the curve is closed, the average value of S (symbolically $E(S)$) cannot exceed $\sqrt{U^2+V^2}$. This follows from Jensen's inequality and the fact that the square root is a concave function. Now, new tape area shows up at the rate UL, but the searcher cannot possibly examine it faster than the rate $2RS$ at any time or $2RE(S)$ on the average. The ratio of these two rates is therefore an upper bound on the detection probability:

$$P_D \leq \frac{2RE(S)}{UL} \leq \frac{2R}{L}\sqrt{1+\frac{V^2}{U^2}}. \tag{7.6}$$

Of course, P_D must also be smaller than 1.

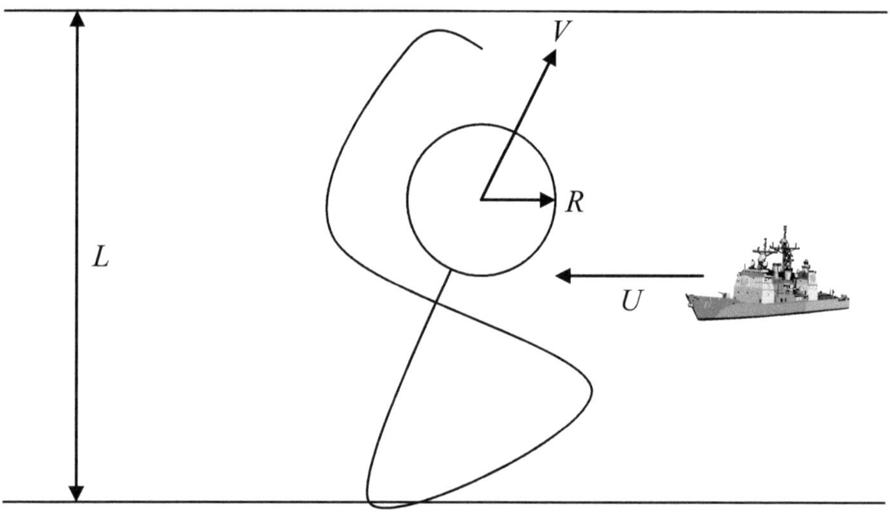

Figure 4: A searcher following a figure-8 type track at speed V hopes to detect a ship moving at speed U as it tries to penetrate a barrier of width L.

Referring to Figure 4, if the searcher were to move entirely in the East–West direction, his average speed relative to the tape, which is moving West at speed U, would be V, the same as if he never moved at all. If the searcher instead moves in the North–South direction, he achieves the preferred dynamically enhanced speed of $\sqrt{U^2+V^2}$, relative to the tape. The trouble is that he must reverse course when he nears the North or South border in order to keep his sensor on the tape, and this course reversal results in wastefully covering parts of the tape that have already been covered. Use of a figure-8 track, instead of a straight back-and-forth track, to some extent achieves a course reversal without double coverage, but only partially. Use of (7.6) amounts to the assumption that the searcher can find some way of achieving dynamic enhancement without suffering much from wasteful coverage of areas outside the tape or wasteful double coverage of areas on the tape. To put it another way, use of (7.6) amounts to pasting the two tape edges together, which would allow the searcher to instantaneously jump from one edge to the other. If the edges were pasted together, the searcher could have dynamic enhancement all the time, without redundant coverage, by continually circling around the tube formed by the pasted tape.

It turns out that the figure-8 class of tracks is large enough that the searcher can usually find some way of nearly achieving the detection probability of (7.6). Even the class of bow-tie tracks (symmetric tracks with four straight edges, two of which are parallel to the channel boundary) is large enough. OEG (1946) considers only two bow-tie tracks, one of which is the special case of back-and-forth patrol. Washburn (1976) generalizes and finds that, in most cases, the bow-tie class is sufficiently large to include a member whose detection probability is close to (7.6).

Example 2: Suppose $U = V$ and $W = L/2$. Formula (7.6) gives 0.707 as an upper bound on detection probability. Exhaustive examination of all bow tie tracks with straight edges where the detection circle just grazes the channel boundary at turns gives a detection probability of about 0.69, achieved with a bow tie whose central legs make an angle of about 81 degrees with the channel boundary. If that angle were 90 degrees, we would have a back-and-forth track with a detection probability of 0.68. A wider selection of tracks, including tracks that slightly overlap the channel boundary at turns, would produce a slightly greater detection probability than 0.69, but surely not greater than 0.707.

7.5 Optimal Distribution of Effort for Stationary Targets

The greater part of search theory does not seek optimal tracks, as we did in the section above. The main reason for this is that, as analytical objects, optimal tracks are hard to deal with. A secondary reason is that most searchers do not want to be told exactly what to do, anyway. These problems may disappear in the future, given the increasing capabilities of computation, robots, and navigation systems. Kierstead and DelBalzo (2003), for example, get encouraging results by applying a genetic algorithm to track optimization. Nonetheless, in this section we follow historical precedent in seeking to give only rough guidance to the searcher in his quest for the target, rather than precise directions about exactly what track to follow. We assume that there is only one constraint on searching, essentially the amount of time available for search. For generality we refer to this constraint as "effort." The problem is then to advise the searcher how he should optimally distribute his effort over the various places where the target might be.

Why not just distribute the effort uniformly? One reason is that the target may be more likely to be in some places than others. Another reason is that the searcher might be more effective in places like meadows than in other places like swamps. Both of these reasons lead to the idea of partitioning the overall search area into "cells" where conditions are uniform. The problem is then one of distributing effort among these cells.

7.5.1 Discrete Effort

Here we assume that effort is measured in discrete quanta called "looks." One look might correspond to an aircraft sortie, to the amount of searching that a dog team can do in 1 day or to an arbitrary time period like 1 hour. The cells are numbered from 1 to n. The geometric arrangement of the cells does not matter because the target is assumed not to move, and because we assume that any effort spent by the searcher in traveling between cells is negligible. We also assume that the searcher has no chance of detection unless he looks in the correct cell; that is, we ignore the possibility that the searcher might find the target in one cell while looking in another. We need two kinds of probabilities:

Definitions: The *occupation probability* for a cell is the probability that the target occupies it, given the history of the search so far. The *overlook probability* for a cell is the probability that a look in the cell will not find the target, even when the target is there. Overlook probabilities are assumed to be given data, not depending on the search plan.

Let p_i be the initial occupation probability for cell i. The corresponding vector $\mathbf{p}$ is the initial distribution for the location of the target. This probability distribution in an input; that is, anyone desiring to use the results of this section will have to quantify the initial uncertainty about target location. Richardson and Stone (1971) recount the accomplishment of this in the search for the lost nuclear submarine Scorpion, where cells corresponded to a fixed grid on the ocean. Given the communications history, the existence of a seamount in the area, and conflicting theories about how the submarine met its demise, each cell in that grid was eventually assigned a probability.

Let q_i be the overlook probability for cell i, and $\mathbf{q}$ the corresponding vector. These overlook probabilities might be obtained from one of the formulas of Section 7.3, or might be direct estimates from experts. They could in principle depend on time, although that is not the case here. We assume that all looks are independent, and so, for example, three looks in cell i have an overlook probability q_i^3. This assumption should not be accepted casually, since its verity depends on the nature of the search in each cell. Suppose that a "look" corresponds to the assignment of a human searcher to a particular grid cell for 4 hours, and suppose that searchers are methodical people who spend their 4 hours making an exhaustive search of half of a cell. The overlook probability for one look is therefore 0.5. What is the overlook probability for two looks? We would like it to be zero, since two looks is enough time for exhaustively searching the whole cell, but the independence assumption would have it be 0.25. The latter might actually be more valid, depending on how the search is organized. Perhaps searcher Al cannot remember what he did the first time when he is once again assigned to look at the cell or perhaps the first look is Al's and the second is Moshe's, and the two do not (or cannot) communicate. This kind of forgetfulness or "friction" might be used to justify the independence assumption. If the two looks are each random searches of the whole area, rather than partial exhaustive searches, then of course the independence assumption is automatically true. In the rest of this section, we assume that the independence assumption holds.

If the total number of looks in cell i is x_i, then it follows from the above assumptions and the theorem of total probability that the overall miss probability is

$$1-P_D=\sum_{i=1}^{n} p_i q_i^{x_i}\ . \tag{7.7}$$

Since $\mathbf{x}=(x_1,\ldots,x_n)$ is under the searcher's control, except for being nonnegative and the requirement that the total number of looks is given, the mathematical problem is to minimize (7.7) with $\mathbf{x}$, subject to the constraint that only a certain number of looks are available. One solution to this problem is the greedy method where each look is assigned to the cell for which the detection probability increases the most. Let Δ_i be the increase when $\mathbf{x}$ is augmented by one additional look in cell i. Then, since all terms in (7.7) remain constant except for the one where x_i is increased,

$$\Delta_i = p_i q_i^{x_i} - p_i q_i^{x_i+1} = p_i(1-q_i)q_i^{x_i}\ . \tag{7.8}$$

Since $q_i^{x_i}$ is initially 1 for all i because $\mathbf{x}$ is initially 0, the first look goes to the cell where $p_i(1-q_i)$ is largest. If the winning cell is cell j, then Δ_j is updated by multiplying by q_j, after which the second look is allocated by again searching for a maximum, and so on until all looks are distributed.

Example 3: Suppose $\mathbf{p}=(0.5, 0.4, 0.1)$ and $\mathbf{q}=(0.9, 0.5, 0)$. Cell 3 might correspond to the proverbial streetlight under which the drunk is tempted to search for his lost keys ($q_3=0$, so detection is certain if the keys are there), but the trouble is that he probably did not lose them there ($p_3=0.1$). Table 1 shows how the first five looks would be allocated by the greedy algorithm, breaking ties by choosing the lower numbered cell. Note that cell 3 is searched, but only on the third look, and that most of the looks go to cell 2. The searcher is so efficient in cell 3 that one look suffices, and so inefficient in cell 1 that effort is better allocated to cell 2. The detection probability for all five looks is $0.2+0.1+0.1+0.05+0.05$, computed by adding up the five selected (bold) Δ's from Table 1 or alternatively by applying (7.7) with $\mathbf{x}=(1,3,1)$. Either way, the best detection probability after 5 looks is 0.5.

look #	winning cell	Δ_1	Δ_2	Δ_3
1	2	0.05	**0.2**	0.1
2	2	0.05	**0.1**	0.1
3	3	0.05	0.05	**0.1**
4	1	**0.05**	0.05	0
5	2	0.045	**0.05**	0

Table 1: Showing how Δ is updated and the winning cell selected for the first five looks of Example 3. The winning cell always corresponds to the largest Δ, shown bold.

The greedy algorithm can also be interpreted differently. One can imagine that the searcher at each stage applies Bayes theorem (Appendix A) to obtain the current distribution of the target's position, given that no detection has occurred yet, and then treats the next look as if it were the first, but with the updated occupation probabilities. Consider the situation after three looks in Example 3. The distribution of the target's position, given that the first three looks fail to detect it, is (0.5, 0.1, 0)/0.6 or (5/6, 1/6, 0). Under the same condition, the detection probabilities for the next look are (1/12, 1/12, 0), so the fourth look can go in either cell 1 or cell 2, as in Table 1. A review of the Bayesian math will reveal that this will always be the case – the looks are allocated in the same sequence under either interpretation. The updated occupation probabilities can be obtained by just dividing out the $(1-q_i)$ factors from Δ_i, and then normalizing so that the probabilities sum to 1. When $(1-q_i)$ is again multiplied back in, we are making (except for the normalization factor) the same comparisons as in Table 1. The updated occupation probabilities are thus a sufficient summary of all past search and might be made the basis of a management information system for the protracted search.

The primary output of most computerized decision aids for search is a display of the most recently updated occupation probability distribution. A recommendation of what to do next might also be made, but the software is always prepared for the user to ignore it. The user makes his decision, observes the result, the occupancy display is updated to account for the new observation, and the cycle repeats. Sheet "Greedy" of *Chapter7.xls* contains an implementation. Terms used on that sheet are taken from terrestrial search and rescue work. The $(1-q_i)$ factors are called PODs (POD for Probability Of Detection) and the input distribution **p** is called POAs (POA for Probability Of Area). Applying Bayes theorem is called "shifting POAs."

The Greedy algorithm described above at no point requires the user to know how much search effort remains. It would be equally applicable if the searcher intended to search until the target is found, whatever number of looks are required, in which case the algorithm would minimize the average total number of looks expended. The Greedy sequence of looks is optimal in practically any sense of the word. This kind of uniform optimality depends on the assumptions made earlier in this section, particularly the assumption that the target does not move.

7.5.2 Continuous Effort

We might measure effort in "looks" even when fractional looks are permitted, but in this section we will simply refer to time, since this is the most common measure of continuous effort. It is convenient to write the overlook probability q_i as $\exp(-\alpha_i)$, where α_i is some nonnegative number (the natural log of $1/q_i$, to be exact). In that case, (7.7) takes the form

$$1 - P_D = \sum_{i=1}^{n} p_i \exp(-\alpha_i x_i). \tag{7.9}$$

The problem of minimizing (7.9) now involves the continuous, nonnegative variables **x**, subject to the constraint $\sum_{i=1}^{n} x_i \le t$, where t is the amount of time available. We can no longer employ the greedy algorithm, since there is no time quantum, but the minimization problem is nonetheless not computationally difficult. There are efficient procedures tailored to the problem (Washburn, 1981), but general nonlinear optimizers such as Excel's Solver will have no difficulty with it. Sheet "Continuous" of *Chapter7.xls* will do the minimization for four cells (Exercise 5).

Equation (7.9) retains the independence assumption of Section 7.5.1. That assumption is correct if the search of every cell is random; in fact, the parameter α for each cell is simply VW/A, as in Section 7.3.2. If the independence assumption is not correct, (7.9) should be replaced by

$$1 - P_D = \sum_{i=1}^{n} p_i f_i(x_i), \tag{7.10}$$

where $f_i(x_i)$ represents the overlook probability when x_i units of effort are applied to cell i. For example, if the search of cell i were according to the inverse cube law, then $f_i(x_i)$ would be an expression involving the cumulative normal distribution. As long as $f_i(x_i)$ is a decreasing, convex function for all cells, efficient techniques for the minimization still exist (Stone, 1975).

7.5.3 Search of a Bivariate Normal Prior

Uncertainty about the target's location often takes the form of a normal distribution in two dimensions. A common reason for this is that several sensors measure bearings to the target. If the sensors make normally distributed errors in measuring bearings at long range, then the resulting position distribution of the target will be bivariate normal. A bivariate normal prior distribution can be illustrated on a map by drawing a standard ellipse, typically the 2-σ ellipse that contains the target with probability 1−exp(−2) = 0.865. Other standards are occasionally used, resulting in larger or smaller displayed ellipses, but any such ellipse is a surrogate for a complete stretched bell-shaped distribution that could be partitioned into cells for purposes of conducting a search. If the search is random, a standard problem results that has been the focus of analysis since World War II. This section summarizes the results.

We suppose that the searcher moves at speed V and sweepwidth W for a time period t. The only important thing about these three quantities is the product, which can be thought of as the amount of area covered by the searcher if overlap is ignored. If the two standard deviations characterizing the ellipse are σ_U and σ_V, we can define a dimensionless measure of effort $z \equiv VWt/(2\pi\sigma_U\sigma_V)$, a kind of coverage ratio. This measure is also defined in Chapter 2, except that the numerator there is na, rather than VWt. The marksman of Chapter 2 is trying to kill the target with n shots, each of which has lethal area a, while our searcher covers area by moving continuously. The marksman and the searcher each face the problem of arranging a given total area to "cover" the target with the largest possible probability.

As in Chapter 2, we distinguish three cases. The exhaustive case corresponds to the optimistic assumption that the total area can be cut and shaped as the searcher wishes, without overlap. It leads to the upper bound (2.18). The other two cases both assume that the search is random, but differ in how the effort is arranged. If tactical simplicity demands that the search effort be distributed uniformly over some region, then the best detection probability is given by the SULR formula (2.20), with the optimal region being the ellipse recorded in Chapter 2. If the random search effort can actually be distributed optimally (this is not a contradiction in terms – think of strewing confetti non-uniformly), then the best detection probability is given by the SOLR formula (2.23). A SOLR search might be implemented by making repeated passes over progressively larger regions. As Figure 4 of Chapter 2 makes clear; however, there is little difference between the SULR and SOLR cases.

There are cases intermediate between exhaustive and random search, as in the next example.

Example 4. Suppose that $\sigma_U = 20$, $\sigma_V = 10$, $V = 100$, $W = 2$, and $t = 8$, so $z = 1.27$. We assume that search must be uniform within some region. Using (2.20), the SULR P_D is 0.46. However, suppose that random search is felt to be too pessimistic an assumption, and that the inverse cube law is expected to apply instead. The idea is to perform a search with uniform track spacing inside some elliptical region, so (7.5) should apply, but what region should be searched? Since no formula for the best region is available, we can only try trial and error. Sheet "BVN" of *Chapter7.xls* finds the best x by exhaustion, where x is the area of the elliptical search region divided by the normalizing constant $2\pi\sigma_U\sigma_V$. The best x turns out to be 1.2. From the definition of x, the best ellipse to search should have radii of $\sqrt{2x}$ standard deviations in each direction. The two diameters of that ellipse are thus 62 and 31. The uniform track spacing within this ellipse should be W/z or

1.57. The expected detection probability is 0.57, significantly higher than the SULR value.

7.6 Moving Targets

If the search process is extended in time, as is often the case, then it may be necessary to allow for motion of the target as the search proceeds. A lost person, for example, can move a significant distance during the daylight hours when searches are usually conducted. If the distance moved during the search time is small compared to the cell dimension, we might proceed as if the target were stationary. In general, however, some method of allowing for target motion is needed.

The significance of target motion depends strongly on what motivates the motion. The target might be cooperative, as when a lost person moves toward a rescuer. In combat situations, however, the target is more likely to move away from the searcher, thus frustrating detection. In fact, one reason for not using active sensors is that the noisy pings will warn the target of the searcher's position, thus enabling avoidance.

Both cooperative and noncooperative behavior depend on the target's early detection of the searcher. It is also possible that the target does not have such information or is moving around with no goal in mind that has anything to do with the possibility of being detected, a type of motion we might call "benign." It is not clear whether benign motion is good or bad for the searcher – an argument can be constructed either way. The target's motion can cause it to find the searcher, as any spider knows. On the other hand, the randomness of target motion is one of the things that can spoil an attempted exhaustive search. Aspects of each argument can be seen in the sections that follow.

7.6.1 Dynamic Enhancement

If the target moves "at random" within the search area at speed U, then it is reasonable to suppose that the angle θ between the target's velocity and the searcher's velocity is equally likely to be anything between 0 and 2π. If the searcher's speed is V, then the relative speed between the two is $\sqrt{U^2+V^2-2UV\cos(\theta)}$. The average relative speed is therefore

$$\tilde{V}=\int_0^{2\pi}\sqrt{U^2+V^2-2UV\cos(\theta)}\,\frac{d\theta}{2\pi}. \tag{7.11}$$

The quantity $\tilde{V}$ is sometimes taken to be a "dynamically enhanced" search speed in the sense that a searcher with speed $\tilde{V}$ will have the same prospects of finding a stationary target as would a searcher with speed V looking for a target moving at speed U. Its use in that sense goes back to at least World War II (OEG, 1946, Section 1.5).

Equation (7.11) is a symmetric function of U and V; it makes no difference whether the searcher moves at 10 knots while the target moves at 2 knots, or vice versa. The enhanced searcher speed is the same (10.1 knots, to be precise) in either case. This observation might cause a target who does not wish to be found to wonder whether movement is actually a good idea.

Although there is no closed form for the integral in (7.11) (it is a complete elliptic integral of the second kind), we can at least give some upper and lower bounds on $\tilde{V}$:

$$\max(U,V) \le \tilde{V} \le \sqrt{U^2+V^2}\,. \tag{7.12}$$

The bounds are closest to each other when U and V differ significantly. The worst case is when U and V are equal (to 1, say), in which case the lower bound is 1, the upper bound is 1.41, and $\tilde{V}$ is actually $4/\pi = 1.27$. Figure 5 shows a graph. See sheet "EquivSpd" of *Chapter7.xls* for similar graphs or to see how to use the VBA function EquivSpd(U,V), which will calculate the dynamically enhanced speed for any arguments.

Example 5. Suppose that a searcher with speed 40 knots is looking for a target with speed 30 knots in a region with area 100 square nautical miles. If the sweepwidth is 0.5 nautical mile, how long will it take for detection? To answer this we first find the dynamically enhanced speed, which turns out to be 63.6 knots. Since the target is moving around randomly, we assume that only random search is possible, and appeal to Section 7.3.2. According to that section, the mean time to detection should be, on the average, $100/(0.5 \times 63.6) = 3.1$ hours.

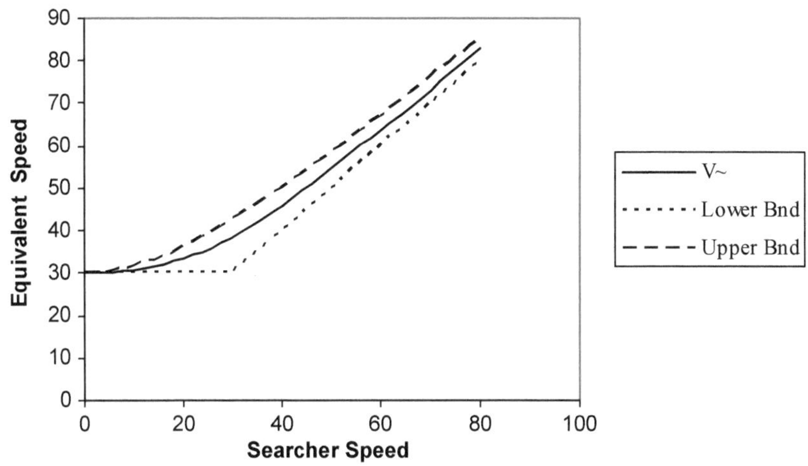

Figure 5: Showing the dynamically enhanced equivalent search speed (*V*~) when the target speed is 30, along with upper and lower bounds.

Although the dynamic enhancement formula has a long history, the authors are not aware of any serious attempt to validate it. Any such attempt would have to be specific about what "random motion" means for both searcher and target. If the target is motivated to avoid detection, then it may very well move slowly even though it is capable of going fast, since significant motion is basically a bad idea for someone who wants to hide. In search games where the target wants to delay detection as long as possible, the target's best motion is known to be almost null, moving just enough to foil any attempt at exhaustive search (Gal, 1980, Chapter 4). If the target of Example 5 moved in that way, the "enhanced" speed would be 40 knots, not 63.6.

7.6.2 Markov Motion

Rather than trying to reduce the situation to an equivalent one involving a stationary target, we might try to extend the theory of Section 7.5 to include moving targets. The simplest concept is to imagine that searcher and target take turns acting. First, the searcher distributes some effort over the cells. If search is unsuccessful, the target may move to a different cell, after which the searcher tries again, and so on. It is not usually realistic to suppose that the target's next position is independent of its current position, so we suppose instead that the target's motion is a Markov chain (Appendix A). This requires that we specify the chain's transition function, in addition to the probability distribution of the target's initial position.

Definition: For a Markov chain defined on a set of cells C, the *transition function* $\Gamma(x,y,t)$ is the probability that a target in cell x at time t will move to cell y at time $t+1$. Such a function must be nonnegative, summing on y to 1 for each $x \in C$, and also for each time t.

A commonly used type of Markov chain is a random walk where the target either moves to a neighboring cell with some specified probability or else remains where it is. The general effect of this is that the distribution of position spreads out over more and more cells as time goes by. Sheet "RandWalk" of *Chapter7.xls* illustrates this for $C=\{1,\ldots,17\}$. That sheet also illustrates the effect of simultaneous searching, but the effect of motion alone can be explored by simply setting the overlook probability to 1.

Example 6: The set of cells is $C=\{1,2,3\}$, with the initial distribution of target position being $\mathbf{p}=(0.51, 0.49, 0)$. After each search, the target increases its cell index by 1 or else does not move if it is already in cell three. In terms of the transition function, we have $\Gamma(1,2,t)=\Gamma(2,3,t)=\Gamma(3,3,t)=1$, for all t, with all other transition probabilities being 0. The overlook probabilities are assumed to be (0,0,1) in the three cells; that is, the target cannot be detected once it gets to cell 3. The overlook probabilities are also 1 if the searcher looks in any cell that does not contain the target. There are two opportunities to look, and the question is where those two looks should be placed. After exhausting all the possibilities, we conclude that both looks should be in cell 2, since this will guarantee detecting the target. If the target is initially located in cell 2, it will be detected on the first look, or on the second look if it is initially located in cell 1.

Let $P(x,t)$ be the probability that the target is located in cell x at time t and is not detected by any of the searches before time t. If normalized to sum on x to 1, this function is the one that a searcher wondering where to look at time t would call "the current distribution of the target's location" and might wish to see displayed to guide his search at time t. Let $q(x,t)$ be the nondetection probability for a look at time t, given that the target is in cell x. This function is determined by the searcher at time t when he decides which cell to look in or possibly how to spread his effort over multiple cells. The formula that advances time is then given by the theorem of total probability:

$$P(y,t+1)=\sum_{x\in C} P(x,t)q(x,t)\Gamma(x,y,t);\ y\in C. \tag{7.13}$$

The only way to get to cell y at time $t+1$, without being detected before $t+1$, is to be in some cell x at time t, without being detected before t, to not be detected at time t, and to move from x to y. That statement is formalized by (7.13). Although Example 6 is simple enough not to require it, (7.13) could be used to reach the same conclusions about the effect of searching twice in cell 2 (Exercise 7). Formula (7.13) is used on sheet "RandWalk" of *Chapter7.xls* to update the distribution of the target's position, accounting for motion as well as the effects of search.

In this paragraph we use the dot (•) notation to emphasize that we are referring to all cells at the stated time, rather than any specific cell. Formula (7.13) might be used in the following way: Examine the initial distribution $P(\bullet, 1)$ and decide on what $q(\bullet, 1)$ should be; that is, decide how to search at time 1. Then use (7.13) to compute $P(\bullet, 2)$ and inspect that function before determining how to search at time 2, thus determining $q(\bullet, 2)$, and so on. In selecting $q(\bullet, t)$ at each time, we might choose the search plan that minimizes the immediate nondetection probability, which is given by (7.13) at time t with the transition function factor $\Gamma(x,y,t)$ omitted. Such a search is myopically optimal in the sense that it maximizes the immediate detection probability at each time, given the failure of all previous searches. Finding a myopically optimal search is computationally easy because it amounts to solving a succession of static search problems, one at a time.

A myopic search will usually come close to maximizing the overall detection probability, but can make mistakes. Applied to Example 6, it would first search cell 1 and then not know what to do for the second look, since all three alternatives result in 0 for the immediate detection probability on the second look (see Exercise 7). The overall detection probability after two looks would be 0.51, the same as the detection probability after one look. There is a better way to utilize two looks, as explained in Example 6. In fact, that example has been deliberately designed to make myopic search look nearly as bad as possible. Myopia is usually much closer to being optimal than in this example, but the fact is that the global optimality property that we found in Section 7.5 disappears when the target moves.

There is a technique – the FAB (Forward And Backward) algorithm – designed to find the globally optimal distribution of effort when the target moves. The main idea is to introduce into (7.13) another factor that represents the probability of nondetection in the future, so that the sum is the nondetection probability over all time. The FAB algorithm is iterative and may converge to an optimal solution only in the limit. Nonetheless, the speed of modern computers makes applications feasible. See Brown (1980) or Washburn (2002) for a complete description of the algorithm.

It should be said that the turn-taking models considered in this section ignore the possibility that the target might be detected in one of those turns where the target's location changes, and as such are probably unreliable models for targets that move a lot. Recall that the dynamically enhanced search speed of Section 7.6.1 is minimized when the target is stationary, an indication that targets who do not want to be found should not move much. In contrast, if this section's reluctant target were given his choice of Markov transition functions, his favorite function in many cases would make his next location equally likely to be in any cell, which implies a lot of motion. These are opposite trends, and the contrast ought to be worrisome. In practice, the models of this section are probably searcher-pessimistic, especially for targets that move a lot. Unfortunately, there seem to have been no studies of exactly what "a lot" means.

7.6.3 Evasive Targets

As mentioned earlier, the success of a search operation for a moving target depends on the target's motivation. Three clear cases can be distinguished:

- The target may desire detection. There is a growing literature on this topic. It is important to the search and rescue community, but we will not delve into it here.
- The target may be indifferent to detection or unaware that it is being searched for. This is certainly the case for inanimate objects and is sometimes true for fish, humans, and submarines. Sections 7.6.1 and 7.6.2 apply to the motion of such objects.
- The target may know that it is being searched for and take measures to avoid being found. This is often the appropriate assumption in combat models. This kind of evasive target is the current subject.

Evasive targets that lack information about the searcher's activities may or may not be able to effectively use their capability for movement. Consider the "Princess and Monster" game, which starts when two players are randomly located in the unit disk. The Monster moves at speed 1 until it comes within a given small capture distance r of the Princess, after which the game ends. The Princess can also move, but she does so blindly because she cannot detect the Monster. Eventual detection by the Monster is inevitable, but the Princess would still like to delay it as much as possible. The solution of this difficult game is known (Lalley and Robbins, 1987). An optimal tactic for the Monster is diffuse reflection, as defined in Section 7.3.2. The Princess' optimal tactic has her moving, but just enough to foil any attempt on the part of the Monster to search exhaustively. The net effect of all this is that the Monster in effect performs a random search of the disk. The Princess' capability for motion is of very little use to her – she cannot do things like "go the other way" because she does not know the Monster's location. All she can do is change a potentially exhaustive search into a random one.

One modification of the Princess and Monster game has her knowing the direction to the Monster. This modification is sometimes militarily realistic, since passive interception of active radar or sonar signals reveals precisely that kind of information, usually at long ranges compared to the searcher's detection radius. The Princess will still be captured, but it will take a lot longer. In a similar game played in a rectangle, Washburn (2002, Chapter 2) finds that giving the Princess a speed of only 20% of the searcher's speed results in increasing the mean time to detection by about 40% over what it would be if the Princess had no directional information. This kind of result is part of conventional military wisdom – it is widely recognized that the use of active sensors in the pursuit of an evasive target, while sometimes necessary, gives away important information. Washburn's results were obtained experimentally by having officer–students at the Naval Postgraduate School play the game repeatedly. There is no known analytic solution of this game.

Another famous problem with an evasive target is the Flaming Datum problem. The name comes from a situation where a submarine has just torpedoed a ship, thereby revealing its own position to pursuing ships or aircraft. The burning ship marks a place where the submarine once was, hence the name. An abstract version with a pursuer and evader has five parameters:

- U is the evader's maximum speed (we assume unlimited endurance)
- V is the pursuer's speed
- W is the pursuer's sweepwidth
- τ is the time after the initiating event when the pursuer arrives at the datum, the "time late"
- t is the amount of time that the pursuer spends searching

The exact solution of this problem as a game is again unknown, but, if we assume that the pursuer at all times searches randomly within the gradually expanding circle that represents the evader's farthest distance from the datum, the detection probability can be shown to be (Washburn, 2002, Chapter 2)

$$P_D(t) = 1 - \exp\left(-\frac{VW}{\pi U^2}\left(\frac{1}{\tau} - \frac{1}{\tau + t}\right)\right) \tag{7.14}$$

In this case, results are very sensitive to the evader's speed, which is squared in (7.14). The evader can use his speed effectively to become more and more lost in the two-dimensional plane, whereas the Princess, no matter what her speed, is not allowed to leave the unit disk. Generalizations of (7.14) are considered by Hohzaki and Washburn (2001).

Example 7: Suppose $(U, V, W, \tau) = (20$ kt, 100 kt, 4 nm, 1 h$)$. After searching for 4 h, the detection probability is $P_D(4) = 0.225$. Even in the limit as t approaches infinity, the detection probability is only 0.273. On the other hand, if U is 15 kt instead of 20 kt, the detection probability after 4 h rises to 0.364. Given the choice of infinite endurance or of dealing with a slower evader, the pursuer would far prefer the slower evader. See Exercise 10.

As was mentioned above, neither the modified Princess and Monster game nor the Flaming Datum problem have been rigorously solved as two-person zero-sum games. This regrettable situation can also be expected in other situations where an evasive target can move as the game proceeds, especially if the target benefits from information about the searcher. As analytic problems, most of them are too complicated to solve exactly. This is unfortunate, since game theory is the natural perspective. The analytic problems simplify if the evasive target cannot move once it chooses its initial location. Chapter 9 includes an example.

7.7 Further Reading

Most detection devices, including human eyes and ears, call "target" only if the received signal is stronger than some threshold. Since signals can exceed a threshold even when no target is present, there is a seeming necessity to consider false alarms along with detections. The trade-off between the two can be summarized in a receiver operating characteristic (ROC) curve that shows the false alarm probability that is implied by any given detection probability. The impact of a false alarm ranges from the time delay required to determine that the contact is not truly a target to the 290 lives lost when the USS Vincennes mistakenly shot down an Iranian airliner in 1988, thinking it was an attacking warplane. The omission of all discussion of such things in the sections above is typical of the search theory literature, but there are a few exceptions. Hohzaki (2007) considers a search game where false alarms consume the searcher's limited time. Also see Stone (1975, Chapter 6) and Washburn (2002, Chapter 10). Another important effect of false alarms is their consumption of munitions. In this book, the UCAV models of Chapter 9 include this effect, which can be important.

While search games such as the Princess and Monster game are difficult mathematical objects, there is much to be learned from their study. The excellent search survey by Benkoski et al. (1991) includes several references to the subject, as well as additional papers involving false alarms.

In practice, searching for a target should be halted if the target is not found after a certain time limit, even if continuation is feasible. In search and rescue operations, this issue is sometimes handled by initially including an unsearchable cell called ROW (Rest Of World). As other cells are searched unsuccessfully, Bayes theorem will gradually move probability into ROW, and search stops when the ROW probability becomes so high that future success is unlikely. Of course, "high" is a relative term, and there is always a trade-off between the value of finding the target and the cost of continuing the search. See Stone (1975, Chapter 5) for relevant analytic optimization methods.

The sections above implicitly assume that the object of search is a single target. As long as all targets are statistically identical and independent, much remains the same if the number of targets is unknown or random. The detection probability after a fixed time is the same for all targets, for example. However, one does have to distinguish whether sampling is with or without replacement; that is, can the searcher remember whether a found target has been found before? If the searcher can remember and knows how many targets are present, then one possible goal of search is to find all of the targets. These issues arise in the UCAV part of Chapter 9. Many other interesting extensions can be found in books devoted entirely to search, for example, Stone (1975) or Washburn (2002).

Exercises

(1) Use (7.2) and (7.3) to prove that the mean time to detection in an exhaustive search is $A/(2VW)$.

(2) In Section 7.3.1, a claim is made that the time to detection lacks memory. Prove that claim by evaluating the conditional probability that it is based on. Ans. The conditional probability should be $\exp(-\lambda x)$, regardless of t.

(3) Formula (7.4) involves a limit for large n. Demonstrate that the limit holds by making a spreadsheet that has a column for $n = 1,2,3\ldots$. The detection probability should be equivalent to (7.3) when $n = 1$ or to (7.4) when n is large. For simplicity, set $VW/A = 1$.

(4) Suppose that five searchers are available to construct a barrier across a channel with width 20 miles, and that each searcher has a definite detection range of 1 mile. Also suppose that $U = V$.
(a) If the five searchers are able to patrol coherently in such a manner that they are jointly equivalent to one searcher with detection radius 5 miles, what will be the detection probability? Approximate it using (7.6).
(b) Same as above except that the searchers patrol independently at different places in the channel, so that each has an independent chance of detecting any transitors.
Ans. 0.71 in (a) and (using (7.6) again, but powering up the result for one searcher) 0.53 in (b). One big chance is better than five little ones.

(5) Use sheet "Continuous" of *Chapter7.xls* to solve the following four-cell problem, in each cell of which a random search is conducted. The table below gives the data for the four cells, with p_i being the probability that the target is located in the cell. A total of 8 h are available for searching.

Cell number	p_i	V_i (speed)	W_i(sweepwidth)	A_i(cell area)
1	0.1	50	10	5000
2	0.2	100	10	5000
3	0.3	200	10	8000
4	0.4	50	20	4000

Ans: First calculate $\boldsymbol{\alpha} = (0.1, 0.2, 0.25, 0.25)$. After minimizing (7.9), the optimal $\mathbf{x}$ is (0, 0.7, 3.1, 4.2), with detection probability 0.45. Given the limited amount of time available, it is best to gamble that the target is not in cell 1, spreading the effort unevenly over the other three cells.

(6) Example 4 utilizes sheet "BVN" of *Chapter7.xls* to solve a problem involving inverse cube law search. It is not necessary to use that brute force technique to solve the corresponding random search problem, since formula

(2.20) gives the best detection probability analytically, but do it anyway. Modify the sheet so that random search replaces inverse cube law search, thus verifying that (2.20) is correct. In Example 4, the best ellipse to search had diameters of 62 and 31. What are the best diameters if search is random? Ans: The best detection probability with random search should be 0.46. Correct to the nearest tenth, the best x is 1.1, which corresponds to an ellipse that is 30 by 59.

(7) Show that the myopic solution to Example 6 is as stated in Section 7.6.2. Assign the first look to cell 1, since $P(1,1) = 0.51$ is the largest of the three probabilities. This determines $q(\bullet, 1)$, and therefore $P(\bullet, 2)$. Assign the second look using the same principle, compute $P(\bullet, 3)$, and then sum $P(\bullet, 3)$ to obtain the probability that the target survives the first two looks. The transition function at time 1 is given in Example 6, and the transition function at time 2 is the same.
Ans. You should find that $P(\bullet, 3) = (0, 0, 0.49)$, which sums to 0.49.

(8) The same as Exercise 7, but this time search cell 2 first in spite of the fact that doing so does not maximize the immediate detection probability. Then search myopically after that. You should find that $P(\bullet, 3)=(0,0,0)$.

(9) Sheet "RandWalk" of *Chapter7.xls* illustrates the effect of repeatedly searching cell 6 while a target moves according to a random walk over 17 cells.
   a. Experiment with the sheet. If the overlook probability is 1, the target should always survive and the distribution of position at time 30 should be nearly a bell-shaped normal distribution (the central limit theorem is operating). If the overlook probability is 0, cell 6 should act as a barrier that prevents any probability from getting into lower numbered cells from higher numbered cells. Predict what would happen if the diffusion probability were 0 and confirm your prediction using the sheet.
   b. Study the formulas in the columns for cells 5, 6, and 7. They all involve the overlook probability, whereas cells in other columns do not. Once you understand the formulas, modify the columns for cells 11, 12, and 13 to reflect a search with the input overlook probability in cell 12 at every time. Have you done this correctly? A verification test would be to input a symmetric initial distribution. If the distribution at time 30 is not symmetric, there is an error somewhere.

(10) Sheet "FlamDat" of *Chapter7.xls* includes a partially implemented graph of detection probability versus time according to formula 7.14. Complete the sheet by providing the missing formula and verify that the claims made in Example 7 are correct. Hint: If you write the formula with correct absolute and relative references, you will only need to type it once. In Excel™, the number π is Pi(), a built-in function with no arguments.

Chapter 8
Mine Warfare

The US Navy has lost control of the sea to a nation without a Navy, using pre-World War I weapons laid by vessels that were utilized at the time of the birth of Christ.

RADM Alan Smith (1950)

8.1 Introduction

In 1950 during the Korean War, a minefield in Wonsan harbor inspired the opening quote by Admiral Smith. That minefield delayed the planned landing by over a week, while 250 ships steamed back and forth outside the harbor. The US Navy lost four minesweepers in the process of clearing it, and several other ships were also sunk or damaged. Naval mines were first used effectively in the Russo-Japanese war of 1904, where they were decisive in spite of being primitive contact mines. With improved sensors, naval mines have been used effectively in every significant naval conflict since then (Hartmann, 1979). Most of the damage done to the US Navy since World War II has been due to mines (National Research Council, 2001).

Mines are also becoming increasingly threatening on land. Earth's mid-latitudes are populated by an estimated 100 million antipersonnel land mines (Figure 1) left over from various wars. Thousands of innocents have been killed or injured by these mines, to the point where a group of nations has agreed in the Ottawa treaty of 1997 to cease stockpiling and using antipersonnel land mines (Keeley, 2003). This group of nations does not include the major mine producers, however, so continued use of land mines can be expected. The reluctance of many nations to abandon use of such mines is directly related to their low cost and high effectiveness. Very low cost land mines in the form of improvised explosive devices (IEDs) have proved to be effective against road traffic.

What is a mine? The distinguishing features are that a mine is stationary once laid, that its location is concealed, and that it destroys itself in the process of attacking its enemy. A spider fails to be a mine only because it lacks the latter property. Like spiders, mines rely for their effectiveness on the enemy's need to move. Since mines do not themselves move, and since they do not need to reload after detonation, mines can be simple and inexpensive. Since they are inexpensive, they can achieve effectiveness through replication. We generally analyze minefields, rather than individual mines.

A. Washburn, M. Kress, *Combat Modeling*, International Series in Operations Research & Management Science 134, DOI 10.1007/978-1-4419-0790-5_8,

Figure 1: A landmine being cleared by hand. To prevent clearance, some landmines are built to detonate if they are even tilted.

Computer models have been employed for decades in analyzing minefields, both for purposes of planning minefields and for clearing them. Mine warfare is a game of measure and countermeasure in the face of uncertainty, so some of these models are rather complex, but we begin with some relatively simple ones in Section 8.2. Sections 8.3 and 8.4 deal with minefield planning and minefield clearing, respectively. Since planning is often conducted in the expectation of clearance, models based on game theory are applicable. Section 8.5 considers some of the possibilities.

8.2 Simple Minefield Models

Large-scale models of warfare are generally not built to study the details of minefield construction and countermeasures, but still need to represent mine warfare in some simple manner. The object is to retain the essence of mine warfare without including too many details, databases, or megaflops.

The simplest model would be to simply declare a certain region out of bounds "because there are mines in there," perhaps enforced by the rule that anyone who enters the area is automatically killed. While this model gets the idea across, it seriously overstates minefield effectiveness. An individual mine generally controls so little area that most real minefields are more likely to let an intruder pass safely than to kill him. In addition, the effectiveness of a real minefield will generally decrease with time because the intruder is usually able to prevent replenishment. Thus, even though the simple "permission" model does have its uses and charms, it will not be useful for minefield planning or clearance.

A slightly more complex model, but still a simple one, would be to imagine that N mines are scattered independently at random within a region of area A, and that each of these mines has a fixed lethal radius R. Any intruder who comes within R of a mine will be killed and will thereby be rendered incapable (we assume) of actuating any further mines. Suppose an intruder moves a distance D within the minefield. This movement defines an area of size $2RD$ that must not contain a mine if the intruder is to survive. The probability that any particular mine lies within this area is $t = 2RD/A$; t is the "threat" per mine. If there are N mines located independently in the area, the threat from the minefield itself is $T = 1 - (1 - t)^N$. This is an example of "powering up," as introduced in Chapter 2. With probability T, the intruder is sunk and N decreases by one. Otherwise, the intruder survives and N does not change. The number of mines cannot decrease by more than one because a sunk intruder is assumed to actuate no further mines. We will call this model the ENWGS model because it is used in the ENhanced Naval Wargaming System (Wagner et al., 1999).

The ENWGS model fixes the defects of the permission model without becoming overly complicated. If realistic numbers are substituted for t and N, the minefield threat T will be realistically small. Barring replenishment with more mines, the threat will also decrease with time as N decreases.

A slight generalization of the ENWGS formula has been used at times in minefield planning, where the crucial question is how many mines are required to make the initial threat of the minefield be sufficiently large. The generalization is to replace $2R$ by W, the sweepwidth of the mine for intruders (see Section 7.2). Suppose that the minefield is a rectangle with width B, assumed large compared to W, and that each intruder transits the length L of the minefield. Then $t=W/B$, and we have the formula for simple initial threat:

$$SIT = 1 - (1 - W / B)^N = 1 - (1 - WL / A)^N. \tag{8.1}$$

Here N is the initial number of mines and the minefield area A is just BL. Using (8.1) we could, for example, compute the number of mines required to make SIT be at least (say) 0.1 (Exercises 1 and 2).

The ENWGS model might be described as a Markov chain (Appendix A) where the state is the number of mines remaining and a transition corresponds to a penetration attempt. With each transition, the state either stays the same (successful penetration) or decreases by one (casualty). The probability of the latter is given by (8.1).

8.3 The Uncountered Minefield Planning Model (UMPM)

The main questions in minefield planning concern the number and type of mines to be employed. There is something to be said for employing a variety of mine types, particularly in situations where minesweeping may precede intrusion, but in this section we consider minefields consisting of only one type of mine. For any specific type, the main problem is to determine the number that is needed.

Although *SIT* is relevant, minefield planners often consider other measures that involve intruders after the first. The "threat profile" is simply a graph of the threat (probability of not making it through the minefield) to intruders 1, 2, 3,… in sequence, with *SIT* being the first of those numbers. The threat profile is usually a decreasing function as long as the minefield cannot be replenished. One can also imagine a group of intruders attempting penetration, and consider *X*, the number that is lost in the process. The probability mass function of *X* can be graphed as the "casualty distribution," and summarized by the expected value *E*(*X*). The ENWGS model is capable of predicting either the casualty distribution or the threat profile (Exercise 3).

While it is an improvement over a simple permission model, the ENWGS model still has some features that make it unattractive as a minefield planning tool. The chief of these is that it does not deal realistically with channelization. Minefields are usually planned in considerable uncertainty about where intruders will travel, so minefields tend to be much wider than the sweepwidth of an individual mine. A reasonable countermeasure on the part of the intruders is to channelize, by which we mean that all intruders try to follow each other through the minefield. In this way only a comparatively narrow channel through the minefield is ever tested, and mines outside this channel are automatically rendered ineffective through lack of opportunity. This tactic does not affect the threat to the first intruder, but the ENWGS model also predicts the threat to intruders after the first.

In extreme cases, ENWGS predictions can be seriously wrong. For example, suppose that we have cookie-cutter mines and intruders that can perfectly follow each other. If the first intruder makes it safely through the minefield, then he leaves a clear channel behind him that is guaranteed not to have any mines in it, so all intruders after the first will also make it through the minefield safely if they follow him. The threat to the second intruder will be zero if the first makes it through safely. The ENWGS model, on the other hand, predicts that the threat to the second intruder is still *SIT* if the first makes it through safely, since *N* is not changed by the first intruder. Clearly, if our minefield model is to give realistic estimates of the threat profile in the face of channelization, then ENWGS must be modified.

The problem with the ENWGS model is that it makes an independence assumption that is falsified by channelization, an example of the sometimes pernicious effects of the Universal Independence modeling habit described in Chapter 1. For mines to act independently, either the intruders must be in independent locations relative to the mines or the mines must move around between intruders. The first possibility is false because of channelization, and the second is false because remaining stationary is a fundamental mine property. In the US Navy's Uncountered Minefield Planning Model (UMPM), the independence assumption is replaced by one of *conditional* independence. Specifically, in UMPM we assume that the effects of a given mine on a succession of intruders are independent *given* the location of the mine relative to the channel centerline.

Let x be the distance between a mine and the channel centerline, and let $A(x)$ be the probability that the mine detonates when an intruder follows the centerline through the minefield (we are ignoring the possibility of navigation errors). In the terminology of Chapter 7, x is a lateral range and $A(x)$ is a lateral range curve, but we will call $A(x)$ an actuation curve in the present application. The area under the actuation curve is the sweepwidth W. Figure 2 shows half of a typical actuation curve. The important point to note is that, from the viewpoint of a minefield planner, actuation curves are regrettably "sloppy" in the sense that results are hard to predict compared to a cookie-cutter curve. Imagine that the mine's lethal radius is 40. Sometimes mines will detonate at distances larger than 40 (wasted fires), and sometimes mines will fail to detonate at distances smaller than 40 ("wasted opportunities"). The ENWGS model acknowledges neither kind of waste, but the UMPM model deals with both by using the entire actuation curve in making its calculations, rather than just the sweepwidth W.

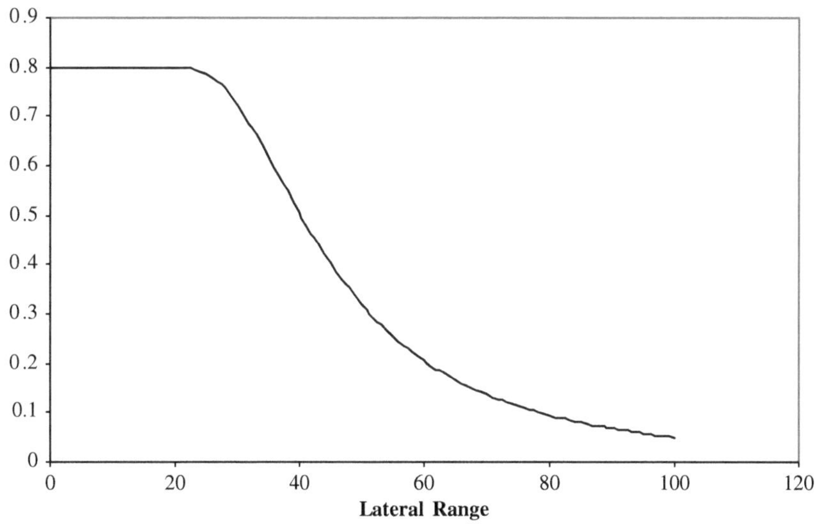

Figure 2. Actuation probability is plotted versus positive lateral ranges. The left-hand side of the actuation curve is symmetric.

Given our conditional independence assumption, the probability that one of n intruders who follow the same channel centerline will detonate a mine at lateral range x can be obtained by powering up the actuation probability:

$$R_n(x) = 1 - (1 - A(x))^n. \qquad (8.2)$$

A minefield planner will also be uncertain about the effects of a detonation, but less so than with actuation. For simplicity, suppose that there is some definite range R such that a mine will be lethal if and only if the lateral range is smaller than R when the mine detonates. If the minefield has width B, and if the channel is perpendicular to this dimension, and if the mine is located at random in the minefield, then the probability that one of n intruders is actually a victim of the mine is

$$R_n{}^* = (1/B)\int_{-R}^{R} R_n(x)dx;\; n > 0. \tag{8.3}$$

In (8.3), the reason for dividing by B is that x is assumed to be uniformly distributed over the width of the minefield. The formula is written as if the channel centerline were the center of the minefield, but is actually valid for any centerline as long as either R or W is small compared to B. The first step in UMPM is to compute the probabilities $R_n{}^*$ using (8.3), for n ranging from 1 up to S, the maximum number of transitors expected.

Once the probabilities $R_n{}^*$ are known, we can generate threat profiles and casualty distributions. Suppose a group of S intruders attempts to penetrate the minefield. With no loss of generality, imagine that the group all attempt to penetrate simultaneously, so that the number of intruders remaining in the group can serve as the state of a Markov chain. If the group has n intruders remaining when it encounters a mine, it loses a member with probability $R_n{}^*$. It is not possible to lose more than one member because one mine cannot kill multiple intruders, so the only other possibility is that the number of intruders does not change. If there are initially S intruders, then the transition matrix is an $(S+1) \times (S+1)$ matrix with at most two positive probabilities in each row. For example, if $S = 3$, then

$$P = \begin{bmatrix} 1 & 0 & 0 & 0 \\ R_1{}^* & 1-R_1{}^* & 0 & 0 \\ 0 & R_2{}^* & 1-R_2{}^* & 0 \\ 0 & 0 & R_3{}^* & 1-R_3{}^* \end{bmatrix}. \tag{8.4}$$

The last row of P^m, the m-th power of the matrix P, will show the probabilities of being in the various states after the group of intruders passes m mines. This row is effectively the casualty distribution. The threat profile can also be obtained from the same matrix, as will be illustrated by example.

Example 1: Suppose that $R_n{}^* = 0.5, 0.6$, and 0.8 for $n = 1,2,3$, and that the minefield initially contains four mines. What is the casualty distribution if three intruders attempt to penetrate the minefield? We first raise the transition matrix to the fourth power to obtain

$$P^4 = \begin{bmatrix} 1 & 0 & 0 & 0 \\ 0.9375 & 0.0625 & 0 & 0 \\ 0.7530 & 0.2214 & 0.0256 & 0 \\ 0.5040 & 0.3894 & 0.0960 & 0.0016 \end{bmatrix}.$$

The last row shows the probability that 0, 1, 2, or all 3 intruders remain alive, which is the desired casualty distribution in reverse order – the probability of no casualties out of three is 0.0016 and the probability that all three intruders are killed is 0.5040. Other information is obtainable with a little more effort. The average number of casualties out of three intruders (call it μ_3) can be found by weighting the probabilities, summing, and subtracting from 3. In this case we find $\mu_3 = 2.4048$. The second and third rows are essentially the casualty distributions if only 1 or 2 intruders enter initially, and can be similarly used to find $\mu_1 = 0.9375$ and $\mu_2 = 1.7274$. The threat to the *i*th intruder is in general $t_i = \mu_i - \mu_{i-1}$ for $i \geq 1$. In this case we find $t_1 = 0.9375$, $t_2 = 0.7899$, and $t_3 = 0.6774$.

The main computational tasks in UMPM are the initial computation of R_n* using (8.3), and the subsequent raising of P to the appropriate power. Once that is done, the casualty distribution, the threat profile, and the mean numbers of casualties are all available with a little more arithmetic. Sheet "UMPM" of *Chapter8.xls* is an implementation for a particular class of actuation curves of the form $A(x) = A\exp(-(C/x)^D)$, all of which look more or less like Figure 2. The parameter A is the maximum possible actuation probability, C is a scale parameter, and D is a shape parameter. When D is small, the actuation curve is sloppy. As D becomes large, the actuation probability drops suddenly from A to 0 as the lateral range increases through C, with the cookie-cutter curve being the limiting version for large D. The only other inputs on the UMPM sheet are the lethal range R, the minefield width B, and the number of mines. The maximum number of intruders is always taken to be $S = 10$. The probabilities R_n* are computed directly on the spreadsheet by numerical integration. Pressing the command button initiates the Markov calculations that ultimately result in a graphically displayed casualty distribution and threat profile.

The UMPM model actually employed by the US Navy differs from sheet "UMPM" in handling actual actuation curves, indefinite damage distances, and intruder navigation errors, but is otherwise similar. Odle (1977) provides the underlying mathematics.

The reader may already have observed that the "uncountered" part of UMPM's name is not exactly appropriate, since the model correctly reflects the results of the assumed channelization countermeasure. Another countermeasure to a minefield is to have the most valuable or vulnerable intruder be the last to enter. This possibility is responsible for the minefield planner's interest in the threat profile, particularly for intruders late in the sequence. One method for increasing the threat to late intruders is the use of probability actuators. When a mine receives signals that it interprets as the presence of a target, it detonates with a probability that is built into the actuation mechanism. Quantitatively, the effect of this can be seen by simply reducing the parameter A in the spreadsheet. The effect will be to decrease the threat to the first intruder, but to increase the threat to late intruders.

A minefield planner usually has multiple concerns that are not easily put on a single scale of utility. UMPM is therefore a descriptive decision aid in the sense that it evaluates a given decision through various graphs and statistics, rather than producing a decision that is "optimal" according to a single criterion. This is true of the US Navy's version of UMPM, as well as sheet "UMPM" of *Chapter8.xls*.

The US Navy's UMPM is not a Monte Carlo simulation, but it could be. Sheet "MonteUMPM" of *Chapter8.xls* includes a command button that performs a Monte Carlo simulation to estimate the casualty distribution. None of the UMPM sheet's numerical integrals are necessary, so the MonteUMPM sheet looks much less cluttered than the UMPM sheet. You are probably using a computer where 10,000 replications can be performed in the blink of an eye, so the computations could certainly be made accurately in the time available for planning a minefield. The threat profile is not calculated, but easily could be. To see the Visual Basic code that performs the simulation, look at subroutine Monte() in the attached module.

The MonteUMPM sheet makes exactly the same assumptions as the UMPM sheet in order to facilitate comparisons between the two. However, the real advantage of Monte Carlo simulation, and our reason for bringing this subject up, is the ease with which alternative assumptions could be implemented. In subroutine Monte(), intruders could be assigned properties such as a type or navigation error, and mines could have properties such as depth, type, sensitivity, or orientation. This easy extensibility is not true of analytic methods such as the Markov chain calculations that lie behind UMPM. As each realistic detail is incorporated, analytic models will be able to avoid excessive complication only by employing the Universal Independence assumption in circumstances where it is false. An intruder's type or navigation error, for example, while random, is not *independently* random for every mine in the minefield. This pervasive temptation to falsely employ UI does not beleaguer Monte Carlo simulations and is one of the main attractions of the technique.

8.4 Minefield Clearance

It takes much longer to clear a minefield than it does to make one, especially if the minefield is designed in the first place to make clearance difficult. Part of the difficulty is that the clearance forces do not know certain minefield characteristics that are concealed by the minefield planner: the minefield location and dimensions, the number and type of each mine employed, the location and sensitivity of each mine, and possibly other mine settings. One consequence of this uncertainty is that minefield clearance is dangerous. Another is that it is difficult to tell exactly when a minefield is "clear" of mines or when to stop clearing.

There are basically three methods for clearing a minefield: destruction, hunting, and sweeping. Destruction is simply applying sufficient lethal force to destroy all the mines, regardless of their type or settings. This is simple in concept, but often expensive in terms of monetary cost and environmental degradation because the lethal force must be applied to every part of the minefield's area. We will not consider it further. With regard to the other two methods,

Definition: One *sweeps* for mines by attempting to cause the mine's sensors to detonate the mine in circumstances where the detonation is harmless. If the mine is located by some means not involving its own sensors, and then either destroyed or avoided, one is instead *hunting*.

Both sweeping and hunting are used in practice. Sweeping is very efficient when it works, since it is not subject to false alarms and mines are disposed of automatically. However, sweeping is subject to mine counter-countermeasures such as sensitivity adjustments, multiple sensors, probability actuators, and increasingly sophisticated signal processing algorithms that can distinguish sweeping signals from signals emitted by real targets. Hunting is vulnerable to none of these, but is affected by false alarms, decoys, and signature reduction. An example of the latter is the occasional burial of mines laid on the bottom of the ocean, which leaves them effective while disguising them from sonar. Hunting also bears the necessity of destroying or otherwise rendering ineffective any mines that are found.

Sweeping and hunting are both essentially search problems in the sense that the object is to "find" all the mines in one manner or another. Minefield clearance might even be treated as an exhaustive search problem – after carefully going over the minefield once, one would declare the minefield to be "clear." However, there are many reasons why exhaustive search might not be possible. In addition to the reasons recounted in Section 7.3.2, in the present context we can add the possibly unknown effects of counter-countermeasures such as probability actuators. Regardless of the amount of sweeping, it is always possible that a mine with a probability actuator is still there and functional. As a result, minefield clearance models generally reject the concept of exhaustion, reporting "clearance level" instead.

Definition: The *clearance level* of a minefield is the probability that a typical mine has been cleared. Equivalently, it is the (average) fraction of the mines that have been cleared. If there are multiple mine types, there may be multiple clearance levels.

Historically, the United States Navy has relied on a pair of clearance level models called NUCEVL (Non Uniform Coverage EVaLuator) and UCPLN (Uniform Coverage PLaNner). The former is descriptive (the sweeping plan is an input, and the clearance level is output), while the latter is prescriptive (the desired clearance level is an input, and the optimal plan among those with constant track spacing is an output). Both of these are incorporated in the Mine Warfare and Environmental Decision Aids Library (MEDAL). They are also utilized in the North

Atlantic Treaty Organization (NATO) decision aid MCM EXERT (Redmayne, 1996), in conjunction with the Decision Aid for Risk Evaluation (DARE, Bryan, 2006). Many other minefield clearance models have been proposed and at times utilized, two of which are COGNIT (McCurdy, 1987) and MIXER (Washburn, 1996). Pollitt (2006) gives a historical review. Some of these models deal with concepts other than clearance level, as should be expected of a subject where there is room for disagreement about goals.

The rest of this section consists mainly of discussing some of the important questions that must be answered before any new model of minefield clearance can be developed. For brevity we will refer only to sweeping in the rest of this section, but we do not mean to exclude hunting. After considering these questions and the differing ways in which the models mentioned above have dealt with them, a prototype minesweeping model will be described in Section 8.4.7.

8.4.1 Are Clearance Forces Vulnerable?

The Wonsan minefield mentioned in the introduction was unusual in its lethality to minesweepers; it is not unusual for minesweeping forces to take no casualties at all in the process of clearing a minefield. This realistic experience is one good reason for not including the possibility of minesweeper casualties in a minesweeping model. Other reasons are that any attempt to do so will force a need for data to support minesweeper vulnerability, and that an additional goal (avoiding minesweeper casualties) will have to be considered. There are thus some good arguments in favor of models where minesweeper casualties are not represented.

On the other hand, the typical real world absence of minesweeper casualties is no accident, but rather the result of carefully considered procedures for clearing minefields safely. If the minesweeping model does not represent even the possibility of minesweeper casualties, then any attempt to optimize tactics might produce tactics that, while efficient in terms of time and clearance level, are dangerous to minesweepers. A modeling dilemma results, the solution to which will depend on context and available data. Among the models mentioned above, NUCEVL, UCPLN, and MCM EXPERT do not include minesweeper casualties, whereas COGNIT and MIXER do.

There is one additional issue if minesweeper casualties are included, and that is the influence of casualties on the ability of the minesweeping force to carry out its plan. The simplest assumption is that all minesweeper casualties are instantly replaced by a miraculous mechanism. Under this assumption, the number of minesweeper casualties is computed and included in any optimization objective, but the minesweeping plan is carried out regardless of casualties. The advantage of this simplification is that analytic models are still possible and are simple enough to permit optimization of tactics. COGNIT makes this assumption, and so does the optimization part of MIXER. The alternative is to include the effects of minesweeper casualties on the minesweeping plan, including the possibility that the plan might have to be prematurely terminated due to lack of equipment. This additional realism is likely to force the model to become descriptive, rather than prescriptive. MIXER includes a Monte Carlo simulation that functions in this mode. There are thus three reasonable stances for a model to take with regard to minesweeper casualties:

- Assume that there are none, as in NUCEVL, UCPLN, and MCM EXPERT.
- Assume that casualties are instantly replaced, as in COGNIT and the optimization part of MIXER.
- Assume that casualties affect clearance, as in the Monte Carlo part of MIXER.

8.4.2 Is Clearance Level a Sufficient Output?

After a clearance operation, a potential intruder might not be satisfied to know only the clearance level. He might instead like to know the threat, the probability that he will become a casualty if he attempts penetration. Any statement about threat requires some knowledge of the number of mines remaining, which in turn requires some knowledge of the number of mines M present in the first place. The trouble is that the clearance forces usually do not know M, even after clearance is finished. We therefore have another dilemma for the modeler. The advantage of restricting effectiveness measures to clearance level is that no knowledge of M is required, and the disadvantage is that potential intruders would rather hear about threat. This dilemma is not present in minefield planning, where M is known and a threat profile is a customary output measure (Section 8.3). However, it is very much present in minefield clearance.

COGNIT and MIXER each require an estimate of M in the form of a probability distribution, prior to computing threat. The need for such a distribution will not be welcomed by the clearance planner, who will often have no idea how many mines are present. An alternative to requiring a user input is to build in a distribution that is not controlled by the user. MEDAL and DARE estimate threat based on the assumption that all numbers of mines are equally likely. NUCEVL and UCPLN confine themselves to clearance level and therefore do not require any assumptions about M.

8.4.3 Is Clearance a Sequential Process?

Information about the minefield is obtained in the process of clearing it, and the decision as to whether to continue clearance or not might depend on results achieved to date. After several days of clearance with no mines found, for example, one might reasonably terminate a clearance operation prematurely. There is an optimal stopping theory that is applicable to such problems, but, to the authors' knowledge, this theory has never been applied to minefield clearance. All of the tactical decision aids mentioned above are "one off" in the sense that optimized clearance effort has a fixed time length.

An intermediate position is to plan fixed-length clearance campaigns, but to at least provide a summary of minefield status at the "end" of clearance in case another fixed-length campaign is required. MIXER, MEDAL, and DARE do this, providing the distribution of the number of mines remaining at the end of clearance.

8.4.4 Are There Multiple Mine Types?

Some models capable of dealing with one type of mine can be easily modified to deal with several. For descriptive, clearance level models such as NUCEVL, it is merely a matter of doing separate calculations for each type, reporting the clearance level for each one. As long as the numbers of mines of each type are independent, the total minefield *SIT* can also be obtained by simply powering up the individual *SIT*s. MIXER considers multiple mine types, including the possibility in the Monte Carlo part that mines of one type can affect the clearance of another type. DARE deals with the possibility that mines might be cleared without revealing their type, but the other models mentioned above do not. Monach and Baker (2006) present a Bayesian method that applies.

8.4.5 Are There Multiple Sweep Types?

In many cases there are a variety of assets available for clearing a minefield. Clearance of maritime minefields has historically been done mainly by manned ships, but several countries are currently using remotely controlled vehicles, and the US Navy also employs helicopters. These different assets usually do not work simultaneously in the same minefield, since there is a danger that one asset will actuate a mine that will damage another. Several questions therefore arise. There is, of course, the division of effort problem – given a fixed time available for clearance, how should it be split between the various assets? If the vehicles are themselves vulnerable, there is also an order-of-entry problem. If helicopters are available, they will typically enter first because they are least vulnerable, but otherwise the best order of entry may not be obvious. These additional considerations greatly complicate the task of constructing an "optimal" minesweeping plan. MIXER alone deals with this issue.

8.4.6 Still More Questions

Is the minefield rectangular? If so, will all of the clearance effort consist of parallel transits? Is it possible that intruders will abandon penetration attempts if initial casualties seem too high? If counter-countermeasures are contemplated, should the situation be thought of as a game, as in Section 8.5 below? All of these questions have been answered in different ways by different models of minefield clearance. Rather than pursue them, we turn instead to the formulation of one possible "new" model.

8.4.7 A Prototype Minesweeping Optimization Model:OptSweep

In this subsection, we provide one set of specific answers to the questions posed above, and develop the implied model. Specifically, we assume

- no minesweeper casualties
- the objective is to minimize *SIT*, the threat to the first intruder
- clearance is one off, rather than sequential, and
- there are multiple mine types and sweep types

Since the objective is to minimize *SIT*, some assumption about the number of mines of type i needs to be made. We assume that the number of mines present is a Poisson random variable with mean α_i for each of the m types of mine present. We also assume that each remaining mine of type i will provide a known threat of t_i to the first intruder. Thus, for each type of mine under consideration, the user must provide two numbers: α_i and t_i.

In clearing the minefield, the user employs x_j sweeps in configuration j, with the variables x_j being the decision variables in this problem. A configuration might be a helicopter towing a sled that is set to actuate magnetic mines or it might be a ship towing an underwater device that cuts the mooring cables of moored mines. It could be 4 hours of grazing by a herd of goats, if the minefield were on land. Each sweep in configuration j is assumed to have a small probability W_{ij} of independently removing each mine of type i, and we assume that there are n sweep configurations available. The average number of times that a mine of type i is removed is then $y_i \equiv \sum_{j=1}^{n} W_{ij} x_j$. Since there are many attempts at removal, each of which succeeds with a small probability, we take the probability that the mine survives all attempts to remove it to be $\exp(-y_i)$, the Poisson probability that the number of removals is 0. The average number of mines that survive all clearance attempts is then $\alpha_i \exp(-y_i)$, and the average number of lethal mines seen by the first intruder is

$$z = \alpha_1 t_1 \exp(-y_1) + \ldots + \alpha_m t_m \exp(-y_m). \tag{8.5}$$

Since the sum of independent Poisson random variables is itself a Poisson random variable, the number of lethal mines seen by the first intruder is itself a Poisson random variable, and *SIT* is $1 - \exp(-z)$. This is the quantity to be minimized, but z can be minimized directly because $1 - \exp(-z)$ is an increasing function of z.

SIT can be made as small as desired by making all of the x_j very large, but we assume that there are resource constraints that prohibit this. Each helicopter sweep consumes helicopter hours, for example, and there might be only so many helicopter hours available within the time allotted for clearance. There might also be constraints on other vehicle types, or on consumable commodities that are available in limited supply, or on the overall time available for clearance. To complete the formulation as a minimization problem, we must express these constraints mathematically.

Let h_{jk} be the amount of resource k consumed by one sweep of type j, and let there be H_k units of resource k available over the minesweeping period. If there are K types of resource available, then the problem of optimally selecting $(x_1,\ldots,x_n)$ reduces to the following optimization problem:

$$\begin{aligned} &\text{minimize } z = \sum_{i=1}^{m} \alpha_i t_i \exp(-y_i) \\ &\text{subject to } y_i = \sum_{j=1}^{n} W_{ij} x_j;\ i = 1,\ldots,m, \\ &\sum_{j=1}^{n} h_{jk} x_j \le H_k;\ k = 1,\ldots,K, \\ &\text{and } x_j \ge 0;\ j = 1,\ldots,n. \end{aligned} \tag{8.6}$$

This problem has linear constraints and a nonlinear, but convex, objective function. As long as the variables **x** and **y** are not restricted to be integers, this is a relatively simple kind of minimization. Sheet "Optsweep" of workbook *Chapter8.xls* uses Excel's Solver to solve the problem of allocating 4 resources to 7 sweep types in order to clear 5 different types of mine. The reader may wish to experiment with it (see Exercise 8).

8.5 Mine Games

The people who manufacture and employ mines are well aware that attempts will be made to clear their minefields. This has lead to a variety of counter-countermeasures designed to make clearance difficult. Mine hunters looking for metal objects can be foiled by making mines nonmetallic or by introducing cheap metallic decoys. Sensors can be used in groups to complicate simultaneous actuation. An anti-tank mine poses little threat to an individual human, but a field of such mines may have occasional antipersonnel mines interspersed to make the anti-tank mines dangerous to locate. Mines can incorporate timers and counters that delay the mine's action, hoping to delay it until right after the expected clearance campaign is over. All of these actions are taken in anticipation of an action by a sentient enemy, so it is natural to apply the theory of TPZS games (Chapter 6). This section examines three of the many possibilities.

8.5.1 The Analytical Countered Minefield Planning Model (ACMPM)

There is a natural tendency for minefields to become less effective with time on account of sweeping and transits by intruders. Mine counters are one possible tactic for reducing this tendency. A mine on count j will detonate when actuated only if $j = 1$; otherwise, each actuation will decrease j by 1 until the mine is finally "ripe" ($j = 1$). By mixing up the counts of the mines, the minefield planner can achieve a minefield that is threatening for late intruders, as well as for early ones. The problem of determining the ideal mixture of counts is a good candidate for a computerized tactical decision aid. We will describe the planning problem in detail for a minefield with only one type of mine.

Suppose that intruders are all alike as far as actuating mines is concerned, but that one of N intruders (the "chief") is more important than the others. It will normally be in the interest of the intruders to send the chief in last, but the object of the minefield planner is to threaten the chief regardless of where he appears in the sequence. If t_n is the threat to the nth intruder, let t be the smallest of all these numbers. Since the chief may be anywhere in the sequence, the object is to make t as large as possible by cleverly setting the counts of the mines. We are in effect considering a game where the intruders move last, since they are assumed to know the count distribution.

Suppose further that every intruder actuates each mine independently with probability A, and that the chief will be damaged with probability D, conditional on actuating a ripe mine. A mine initially set on count j will be ripe just before the nth transit if and only if the first n–1 intruders actuate it exactly j–1 times, a binomial probability. If P_{jn} is this probability, then

$$P_{jn} = \binom{n-1}{j-1} A^{j-1}(1-A)^{n-j}; 1 \le j \le n \le N. \qquad (8.7)$$

Note that $P_{11} = 1$, since the number of combinations of 0 things taken 0 at a time is by definition 1. If the chief is intruder n, he will be damaged by this mine if and only if it is ripe, and if the chief both actuates it and is damaged by it. The probability of this is ADP_{jn}. If x_j is the number of mines set on count j, and if all mines act independently, then the chief's probability of surviving all mines is

$$1-t_n = \prod_{j=1}^{n}(1-ADP_{jn})^{x_j};1 \le n \le N. \tag{8.8}$$

The upper limit of the product in (8.8) is n because mines initially on counts exceeding n cannot threaten the nth intruder. If there are a total of M mines available, then (8.8) is a product of at most M factors, each of which is the probability of surviving one mine.

We can now consider the problem of maximizing t, the minimum of all the numbers t_n, subject to the constraint that the variables x_j must not sum to more than M. See sheet "ACMPM" of *Chapter8.xls* for an implementation. Solutions can be surprising. One might think that there would have to be some mines on high initial counts to guard against the possibility that the chief is late in the sequence. This turns out not to be true when A is substantially smaller than 1. If A is 0.2, for example, it would not be unusual to have a mine initially on count 10 still be on count 10 after several transits. If A is small enough, in fact, the best tactic will be to put all mines on count 1. It is only in situations where actuation is likely, possibly because most intruders are actually minesweepers, that advanced counts become attractive (see Exercise 6).

The Analytical Countered Minefield Planning Model (ACMPM) is a US Navy program used to design countered minefields (Bronowitz and Fennemore, 1975) that avoids some of the artificial assumptions made above. The UI assumption made above that "every intruder actuates each mine independently" is a bad one in the face of channelization, as we pointed out in Section 8.3. ACMPM instead makes the UMPM assumption of *conditional* independence. ACMPM also deals with a variety of mine types simultaneously and includes resource constraints other than simple constraints on the number of mines. In other words, the real ACMPM improves and generalizes the calculations made on sheet "ACMPM" of *Chapter8.xls*.

8.5.2 Triangular Sweeping

In this subsection we exclude all countermeasures other than sweeping. Once the minefield is discovered, the minesweeping forces randomly select a navigation channel through it. Only the navigation channel will be cleared by sweeping, but even within the channel there is a guessing game going on.

A mine that is swept sufficiently often will detonate uselessly, but a mine on a high count may also be useless if it is not swept at all, since it may never ripen until all intruders have passed. The dangerous mines are those with an initial count just slightly greater than the number of sweeps encountered. A kind of guessing game ensues, since the miner would like to guess the number of sweeps so that he can set the mine counter to be slightly larger than that. To make guessing difficult, the sweeper should be unpredictable in the amount of his sweeping. One way of doing this would be to partition the minefield into equally sized subregions, with the number of sweeps depending on the subregion. If the subregions are numbered 0,1,2,…, and if the number of sweeps in each subregion is the same as its index, then we have "triangular minesweeping." Figure 3 shows an example where the navigation channel is broken up into four subregions, all of which must be traversed by any intruder. Triangular minesweeping is optimal in a sense that we will now make precise.

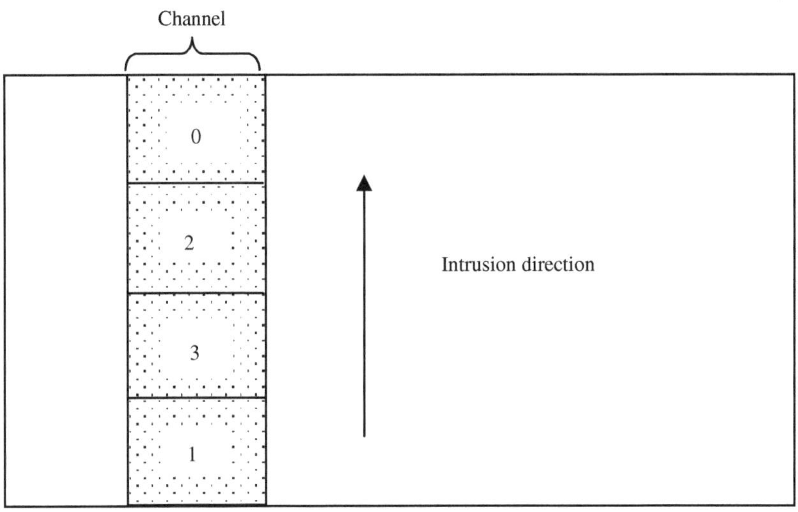

Figure 3: A minefield with a randomly selected navigation channel (speckled) broken up into four equally sized subregions labeled with a random permutation of (0,1,2,3). Each subregion is swept a number of times that is equal to its index.

It is convenient to temporarily change the definition of ripeness so that a mine is ripe when its count is 0, rather than 1. Let x_i be the fraction of mines that are initially set to count i, and let y_i be the fraction of the minefield that is swept i times, both defined for $i \geq 0$. Since the mine counts and the sweeps are determined independently, and since a mine will be ripe after sweeping if and only if the count agrees exactly with the number of sweeps, the probability that a randomly selected mine will be ripe after sweeping is

$$p = \sum_{i=0}^{\infty} x_i y_i. \tag{8.9}$$

The sweeper wants to minimize p with $\mathbf{y}$, and the miner wants to maximize p with

x. We suppose that the average number of sweeps per subregion is limited to y, the coverage ratio (Section 7.3) permitted by the time available for sweeping, and that both sides know y. The sweeper's strategy **y** is therefore subject to the constraint that $\sum_{i=0}^{\infty} iy_i \le y$, in addition to being nonnegative and summing to unity.

For simplicity, suppose that y is a multiple of 0.5, in which case $2y$ is an integer N. Under that supposition, we can find a saddle point for the game. An optimal strategy for the minesweeper is to construct $N+1$ subregions, and to make $y_i = 1/(N+1);\ i = 0,\ldots,N$. Since $\sum_{i=1}^{N} i = N(N+1)/2$, this makes the average number of sweeps per subregion be exactly $N/2$, which is y, so the strategy **y** is feasible. Furthermore, regardless of **x**, as long as **x** is a probability distribution,

$$p = \sum_{i=0}^{\infty} x_i y_i = \sum_{i=0}^{N} x_i /(N+1) = 1/(N+1), \tag{8.10}$$

so the sweeper can guarantee that p is exactly $1/(N+1)$, regardless of how the counters are set.

An optimal strategy for the miner is to set $x_i = \dfrac{2(N-i)}{N(N+1)}$; $i = 0,\ldots, N-1$, and $x_i = 0$ for $i \ge N$. The miner needs to know nothing about subregions in order to do this – the mines can be scattered randomly over the whole minefield. If the miner sets the counts according to **x**, then, regardless of **y**, as long as **y** is a probability distribution whose mean does not exceed $N/2$

$$p = \sum_{i=0}^{\infty} x_i y_i \ge \sum_{i=0}^{\infty} \frac{2(N-i)}{N(N+1)} y_i = \frac{2}{N(N+1)}\Big(N - \sum_{i=0}^{\infty} iy_i\Big) \ge \frac{1}{N+1}. \tag{8.11}$$

We thus see that the miner can guarantee that p is at least $1/(N+1)$. Since both sides can guarantee the same value, the strategies are optimal, and the value of the game is $1/(N+1)$.

Example 2: Suppose that the navigation channel is a square with side 10 km, with area $A = 100$ km^2. The sweeper has two minesweeping vehicles available for $T = 5$ days, each of which is available 15 h per day. The sweeping speed is $V = 10$ km/h, and the sweepwidth is $W = 100$ m. Since there are two sweepers, the coverage ratio is $y = \dfrac{2VWT}{A} = 1.5$ and $N = 3$. The sweeper divides the minefield into four equal parts, secretly and randomly numbered 0, 1, 2, 3, as shown in Figure 3, sweeping i times in part i. The miner sets each mine to count (0, 1, 2) with probability (3/6, 2/6, 1/6). The value of the game is 0.25, so 25% of the mines will be ripe after sweeping. The threat of the resulting minefield depends on the total number of mines used, of course, but the sweeping plan and the mine count setting do not.

The sweeper could get rid of all mines by sweeping all four regions three times, but that would take 10 days, rather than the five that are available.

The value of this game decreases sufficiently slowly with N that the sweeper may have second thoughts about sweeping as a countermeasure. It is tempting to hunt instead, since hunting is not affected by mine counts. Hunting would also spare the sweeper the necessity of explaining to his superiors why he plans to divide an apparently uniform minefield into subregions, all of which are treated differently. There is an explanation (see above), but it is not intuitive.

When the miner uses $\mathbf{x}$, the probability that a mine is ripe after i actuations decreases linearly until it becomes 0 after N actuations. If each mine instead were to contain a probability actuator that detonates the mine with probability A after each actuation, then x_i would be $A(1-A)^i$ for $i \geq 0$, a geometric distribution that also decreases to 0, but only asymptotically. While there is no known sense in which the geometric distribution is optimal, there are still some arguments in its favor. One is its ease of implementation, since all that is required is a probability actuator. Another is the likely falsity of the assumption made above that the miner knows N. If N is actually unknown to the miner, the effect should be to "slop out" the linear decrease into something that looks more geometric.

8.5.3 IED Warfare

Mines are often used by a miner who is about to lose control of the battlespace, if only temporarily. The miner acts, then loses control of the battlespace, then the sweeper acts to clear the mines, and finally intruders pass through the minefield. This is the "one-off" point of view taken in all previous parts of this chapter. In this section we consider a different point of view where the creation and destruction of a minefield proceed simultaneously over an indefinitely long period of time, with neither side controlling the battlespace to the extent that activities by the other side cease. The motivation for this is the kind of warfare conducted on the road network of Iraq in the first years of the current millennium, so we will adopt terminology appropriate to that application. The essence of this problem is that there are many possible places (road segments) for mines, and both sides allocate resources to these segments. Each mine placed on a road segment either detonates as intended by the miner or is found and cleared before it has a chance to do so. In the one-off view there is also the possibility that neither of these events will happen, but that the mine will instead be left over after all intruders have passed. We omit this third possibility on account of the long time periods involved. "Intruders" in this kind of warfare are generally compact groups of vehicles, referred to below as "convoys."

Let i be an index for road segment, and let b_i be the traffic level on segment i, the rate per unit time at which convoys travel over the segment. The probability that a mine on segment i will be actuated by a specific convoy is a_i, which may be less than 1 because of various countermeasures carried by convoys. A mine that actuates (detonates) will be successful with probability c_i, where "success" means significant damage to the convoy. All three of these parameters (a_i, b_i, and c_i) are assumed to be known to both sides, for all segments.

The sweeper has a total number of sweeping teams y that he allocates to the road segments, with y_i being the number allocated to segment i. Each team allocated to segment i removes each mine on that segment at rate β_i, independently of other mines and teams. The sweep effectiveness parameter β_i depends on the speed of the sweeping teams, the length of the road segment, and other parameters that need not concern us – the main point is that the rate of removal is proportional to the number of teams assigned, with β_i being the proportionality constant. The total rate of removal for each mine on the segment, by sweeping teams, is then $\beta_i y_i$. Mines are also removed by convoy traffic when the mines detonate, whether successfully or unsuccessfully, but we have yet to address the rate at which that happens.

We assume that sweeping and convoy traffic are independent Poisson processes, so there is no way for the miner to predict when the next convoy or sweeper will come by. In particular, the miner cannot wait until he observes a sweeping team come by, and then rush out to place a mine on the road before the immediately following convoy arrives.

A small digression is appropriate at this point. One might argue that a better way for the sweepers to operate would be to directly precede each convoy by a sweeping team; that is, one might argue that the sweeping teams should serve as escorts. That way the convoys would always be moving over road segments that have just been cleared. In fact, the assumption of independence between convoy traffic and clearance may strike the reader as an example of the Universal Independence habit that he was warned about in Chapter 1. However, there are some realistic arguments in favor of assuming independence in this case. Among these are

- The sweep teams might be slower than the convoys. Of course, it would always be possible to have the sweep team leave before the convoy, but then there would be an exploitable gap between the two.
- There might not be enough sweeping teams or they might be stationed at different places than where convoys originate. Requiring teams to do escort duty might waste some of their time.
- Sweeping teams might not remove all mines as they pass over a road segment, and therefore cannot guarantee safety even if they do directly precede a traffic unit.

A similar issue arose concerning the escort of convoys in the Battle of the Atlantic in World War II. A given fleet of allied ships could either escort convoys or operate independently of them, prowling the ocean looking for submarines to sink. Both types of tactic were found to be useful at different times in that battle. Here we are ignoring the possibility of escort, assuming that convoys and sweeping teams operate independently.

We now return to an analysis based on the independence assumption. The total rate at which a mine on segment i is removed is $a_i b_i + \beta_i y_i$, the sum of the removal rates due to traffic and sweeping. The proportion of this due to convoys is $\frac{a_i b_i}{a_i b_i + \beta_i y_i}$. This proportion is also the probability that a mine placed on segment i will be detonated by a convoy before it is removed by a sweeping team (see "race of exponentials" in Appendix A). When multiplied by c_i, this is the probability of success for a mine on segment i.

Now let x_i be the probability that a given mine will be placed on segment i, so that the overall probability of success for a mine is $A(\mathbf{x},\mathbf{y}) = \sum_i \frac{a_i b_i c_i x_i}{a_i b_i + \beta_i y_i}$. The miner would like to choose $\mathbf{x}$ to maximize this probability, and the sweeping teams would like to choose $\mathbf{y}$ to minimize it. This is a logistics game (Section 6.2.3), so it has a saddle point whose value can be found by solving a minimization problem. See Exercise 7.

Note that $A(\mathbf{x},\mathbf{y})$ is a probability per mine. Casualties to convoys will be proportional to the number of mines that are placed on the roads, no matter what the clearance forces do. Given the assumptions made above, the clearance forces can only minimize the proportionality constant.

This analysis can be generalized in several ways. For example, one might take the origin–destination traffic levels to be given, rather than the segment traffic levels, and let convoy routing be part of the sweeper's problem. Some of these generalizations are explored in Washburn (2006).

Exercises

(1) Suppose $W = 20$ m and $B = 1000$ m. Using (8.1), how many mines are needed to make *SIT* be at least 0.2?
Ans. $N = 12$.

(2) Make a spreadsheet that will answer questions such as in Exercise 1. The user inputs *W*, *B*, and the desired value of *SIT*, and the required number of mines is calculated.
Hint: *N* can be a ratio of logarithms, rounded up.

(3) Suppose there are three mines in a minefield of width 1000 m, each with a definite lethal radius of 100 m and an actuation probability of 1.0 within that radius. If two intruders attempt penetration of the minefield, what is the probability that both are killed? Compare the ENWGS and UMPM answers.
Ans. Since $B = 1000$ m and $W = 200$ m, the answer for ENWGS is $(0.488)(0.36) = 0.176$. For UMPM, first note that $R_n^* = 0.2$ for all n. After raising the transition matrix to the third power, the answer is 0.104. Alternatively, the probability that the first mine kills an intruder, the second does not, and the third one does is (0.2)(0.8)(0.2). There are two more ways in which both intruders might be killed by three mines, and the sum of all three probabilities is 0.104.

(4) Suppose that a certain type of mine has a definite lethal radius of $R = 100$ m. The mine will always actuate if an intruder comes within 50 m, but the actuation probability is only 0.5 if the closest point of approach is between 50 m and 150 m. The minefield width is $B = 1000$ m. Using (8.3), find a formula for R_n^*.
Ans. $R_n^* = 0.2 - 0.1 \times (0.5)^n$; $n > 0$. Even if a very large number of intruders attempt penetration, the probability that a given mine will kill one of them does not exceed 0.2. Mines outside of a 200 m channel cannot kill anything, even if actuated.

(5) Suppose you are designing a minefield using sheet "UMPM" of *Chapter8.xls*. The minefield width is $B = 500$ m, the shape factor for the mines is 3, the lethal radius is 50 m, and there are 22 mines in total. Those parameters are fixed, but you can change the scale factor up or down from its default value of 30 m by adjusting the sensitivity of the mine's sensor. You can also adjust the actuation probability, but only downward from its default value of 1/3. You are concerned about two quantities: the average number of intruders killed out of 10, and the threat to the last (tenth) transitor. Can you find any way to adjust the two controllable parameters from their default values that improves both measures?

Ans. There are many ways to do it. If you change the scale factor to 35 m and the actuation probability to 0.3, the average number killed improves from 3.59 to 3.71, and the threat to the last transitor improves from 0.135 to 0.143.

(6) Worksheet "ACMPM" of Excel™ workbook *MineWar.xls* implements Equations (8.5) and (8.6) and invites you to determine the best count distribution for a given number of mines. See if you can find the distribution that maximizes the minimum threat when $M = 20$, $N = 10$, $A = 0.4$, and $D = 0.3$. You may wish to take advantage of Excel's Solver feature, which is set up to find the best distribution if the requirement that the number of mines on each count must be an integer is ignored. You may also wish to experiment with smaller values of A such as 0.1, in which case the benefits of advanced counts should be much smaller.

(7) Consider the IED problem described in Section 8.5.3, with five road segments and the data $\mathbf{a} = (0.1, 1, 1, 1, 1)$, $\mathbf{b} = (2, 4, 3, 2, 4)$, $\mathbf{c} = (0.1, 0.2, 0.3, 0.4, 0.5)$, and $\boldsymbol{\beta} = (8, 7, 6, 5, 4)$. If five teams are available, how should they be split among the five segments, and what is the resulting fraction of successful mines? Answer the question by completing the spreadsheet begun on sheet "IED" of *Chapter8.xls*, where the above data are recorded, and possibly consult sheet "Logistic" of *Chapter6.xls*.
Ans. $\mathbf{y} = (0, 0.35, 0.71, 0.89, 3.04)$, and 0.12 of the mines are successful. Most of the sweep teams are assigned to segment 5, which has lots of traffic and unfavorable values for c_5 and β_5. No teams are assigned to segment 1, where the opposite is true and there is in addition a low actuation probability.

(8) Consider the optimization problem formulated on sheet "OptSweep" of *Chapter8.xls*. Study it to make sure that it corresponds to the formulation given in Section 8.4.7. When Solver is employed, the best feasible allocation of sweeps should produce $z = 0.2418$, which corresponds to an *SIT* of 0.2148. If the number of sweeps of each type is forced to be an integer, as Solver permits (you need to add a constraint where the adjustable cells are required to be integers), the minimized z should increase. Exactly how much does it increase? The problem being solved is not of a type where Solver guarantees to find the optimal solution (note Solver's careful characterization of its solution). Can you find a better solution than Solver's?
Ans. In the Solver that comes with Excel™ 2003, z increases to 0.2561 when $\mathbf{x}$ is required to consist of integers, using the optimal non-integer starting point. The "optimizing" $\mathbf{x}$ is (16, 14, 0, 1, 0, 0, 25). The authors do not know whether there is a better solution.

Chapter 9
Unmanned Aerial Vehicles

No operational commander should have to assign a soldier a task that could be done as well by a computer, a remote sensor, or an unmanned airplane.

Richard Perle

9.1 Introduction

This chapter is concerned with *Unmanned Aerial Vehicles* (UAVs). A UAV is a remotely piloted or self-piloted aircraft that can carry a payload of cameras, sensors, communications, and electronic warfare equipment. A UAV may carry also a weapon, in which case it is called an *Unmanned Combat Aerial Vehicle* (UCAV). UCAVs are effective attack weapons. Typical missions of UAVs are surveillance, reconnaissance, target engagement, and fire control for other long-range weapons. UAVs vary in design, size, capabilities, and endurance. Small and lightweight UAVs, with limited endurance and flying range are used for close-range surveillance and reconnaissance by tactical units such as infantry battalions and special operations teams. Larger and heavier UAVs, with higher endurance and longer range, are used for longer reconnaissance missions such as gathering operational-level intelligence. At the far end of the line of UAVs stand very large vehicles that weigh several tons, can endure continuous missions of 24 h and more, and fly at a very high altitude (up to 20 km). The flying range of such UAVs is thousands of km, and their primary use is to provide wide area coverage for strategic information-gathering missions.

In this chapter we consider two types of UAV problems: routing UAVs in reconnaissance or search missions, and evaluating the effectiveness of UCAVs in attack missions. Section 9.2 presents models for optimizing the routes of UAVs in two types of missions, while Section 9.3 describes probability models for evaluating the effectiveness of UCAVs.

UAVs have advantages in not risking a human pilot, and in omitting all the life support systems that a human demands. It does not follow that UAVs are expendable. While small UCAVs are expendable by design, other larger and expensive UAVs, such as the Predator or Global Hawk, cost millions of dollars and are designed to execute many missions. Losing such a UAV has fiscal, as well as tactical consequences. The resulting aversion to loss explains the emphasis on survivability in the next section.

A. Washburn, M. Kress, *Combat Modeling*, International Series in Operations Research & Management Science 134, DOI 10.1007/978-1-4419-0790-5_9,

9.2 Routing a UAV

A mission plan for UAVs concerns three main issues: locating the deployment site of the UAVs' ground control units, routing the vehicles in the area of interest, and scheduling their flights. Three factors affect the UAVs' mission plan:

- objective
- measure of effectiveness (MOE)
- operational, technical, and logistical constraints

The objective of the UAV mission depends on the operational setting and requirements. For example, short-range mini-UAVs, which carry a light payload such as a low-resolution short-range electro-optical sensor, may be operated by an infantry battalion to gather tactical intelligence regarding enemy units at the other side of the hill. Another possible scenario is where a team of Special Operations Forces (SOF) uses such UAVs to search and detect insurgents or terrorists in mountainous areas. Larger, high-endurance UAVs may be used for persistent open-ocean and littoral surveillance of small vessels over extended periods.

The MOE that measures the success of attaining the objective depends on the operational setting. In the SOF scenario a reasonable MOE is the probability of detecting the target (terrorist). In scenarios of persistent surveillance over hostile area the MOE may be the expected duration of the surveillance mission or the probability that the UAV completes its mission unharmed.

Several operational, physical, and logistical constraints must be taken into consideration when planning UAV missions. Operational constraints include no-fly (or high-risk) zones, limited time windows for flying, *deconfliction* among UAVs that share the same airspace (that is, making sure that multiple UAVs do not collide in midair), and limited available deployment sites for ground control units (GCU) that control the flight of the UAV. Physical constraints are derived from limited UAV–GCU communication range, line of sight requirements, possible interference among the various UAV–GCU communication channels, and the limited field of view of a UAV's sensor. Logistical constraints, such as fuel capacity and maintenance requirements, determine the endurance of the UAV.

In this section we consider two types of missions:

- *Intelligence*: The UAV is sent out to investigate a certain object in a certain location.
- *Reconnaissance*: The UAV is on a patrol mission, searching for targets.

9.2.1 Intelligence – Investigating an Object

The mission of the UAV is to fly to a specified destination, investigate a certain object such as a weapon system or an installation, and send back video images of that object.

The Problem. The UAV flies over hostile territory and therefore may be subject to interception by the enemy's air defense weapons. The object of interest may disappear or hide and therefore the mission is time sensitive. In view of these two operational aspects of the mission, we consider two MOEs:

1. *Time*: the time it takes the UAV to reach its destination.
2. *Survivability*: the probability that the UAV reaches its destination unharmed (and thus it is able to execute its mission).

The objectives are to minimize time and maximize survivability, and the problem is how to route the UAV from its base to its destination such that the two objectives are attained. A common and relatively simple way to route UAVs is by specifying a set of *waypoints* that the UAV must follow. These waypoints are entered into the UAV command unit, and the UAV, using navigation devices such as satellite-based Global Positioning System (GPS), moves from one waypoint to another en route to its destination. The first waypoint is the base and the last one is the destination. The possible intermediate waypoints are determined by technical (e.g., line of sight) and operational (e.g., threat zones) considerations. The base, destination and the set of possible intermediate waypoints, along with the segments that connect them, form a graph, as shown in Figure 1. The graph in Figure 1 has 11 nodes – base, destination, and nine intermediate waypoints – and 18 edges, which correspond to operationally feasible flying segments. For example, the UAV can fly from node (waypoint) 2 to node 6, but not to node 4. The problem is to find the best set of intermediate waypoints for the UAV's route, where *best* is defined by the two aforementioned objectives. In this and subsequent models we assume no restrictions regarding the GCU–UAV connection. The GCU of the UAV is located at the base node and it is within range and line of site with all the other nodes. Also, the UAV has no logistical constraints, e.g., the choice of route is not affected by fuel consumption.

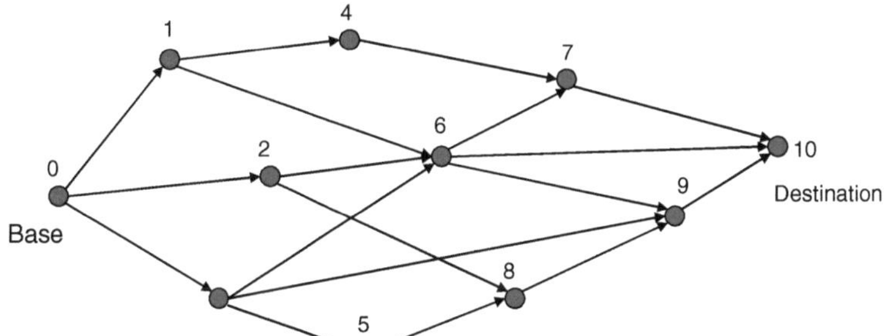

Figure 1: Graph of waypoints.

The Model. Suppose that the set of possible intermediate waypoints comprises W waypoints ($W = 9$ in Figure 1), and we label the *base* node by 0, and the *destination* node by $W + 1$ (node 10 in Figure 1). The directed edge from node i to node j is denoted (i,j). More generally, a route from i_1 to i_2 to … to i_n is denoted $(i_1, i_2, ..., i_n)$. For example, (2, 6, 9) is a route from node 2 to node 9 in Figure 1. Each edge (i,j) in the graph has two associated numbers – the flying time t_{ij} on the corresponding segment, and the probability p_{ij} that the UAV is intercepted while flying on that segment. The parameter t_{ij} is determined by distance, wind conditions, and the nominal velocity of the UAV, and the probability p_{ij} is determined by the locations and capabilities of hostile interception units. We assume spatial independence of the latter; the probability of being intercepted on one edge is independent of the probability of being intercepted on another edge. This independence assumption is reasonable in particular if an enemy air defense unit has limited coverage and is effective with respect to only one edge. Thus, the probability that the UAV reaches its destination safely if it takes route $(i_1, i_2, ..., i_n)$ is $\prod_{k=1}^{n-1}(1 - p_{i_k i_{k+1}})$, which is the *survivability* MOE. The *mission time* MOE for that route is $\sum_{k=1}^{n-1} t_{i_k i_{k+1}}$ – the total travel time. We assume that all the edges are feasible in the sense that for each edge (i,j) $p_{ij} < 1$.

Next we develop an integer linear optimization model (see Appendix B) that determines the optimal route – optimal set of intermediate waypoints – for the UAV. The model addresses the two MOEs, survivability and time, described above.

The optimization problem is a variant of the well-known *shortest path* problem (see e.g., Ahuja et al., 1993).

Variables

The variables in this problem are binary:

$$X_{ij} = \begin{cases} 1 & \text{if the UAV flies over edge } (i,j) \\ 0 & \text{otherwise} \end{cases} .$$

Objective Function

There are two possible ways to model the MOEs in the optimization problem: *time-oriented* model and *survivability-oriented* model. In the time-oriented model the objective is to minimize the travel time to the destination, while maintaining a minimum threshold for the survivability probability. In the time-oriented model the objective is to minimize $\sum_{i=0}^{W} \sum_{j=1}^{W+1} t_{ij} X_{ij}$, the total mission time, while satisfying the minimum probability requirement. This double sum has the same value as the single sum used in defining the mission time MOE above, since the double sum includes the flying time for every edge actually flown, and no others. In the survivability-oriented model the mission time is constrained and the objective is to maximize survivability, under that constraint. It follows that in the survivability oriented model the objective is to maximize $\prod_{i=0}^{W} \prod_{j=1}^{W+1} (1-p_{ij})^{X_{ij}}$. The objective can just as well be the logarithm of that quantity, since the logarithm is an increasing function. Since sums are analytically more convenient than products, we therefore take the objective to be maximizing $\sum_{i=0}^{W} \sum_{j=1}^{W+1} [\ln(1-p_{ij})] X_{ij}$.

Constraints

Both time-oriented and survivability-oriented models share the same constraints that determine feasible flying routes.

$$\sum_{j=1}^{W+1} a_{0j} X_{0j} = 1, \tag{9.1}$$

$$\sum_{k=0}^{W} a_{ki} X_{ki} - \sum_{j=1}^{W+1} a_{ij} X_{ij} = 0, \quad i = 1, \ldots, W, \tag{9.2}$$

$$X_{ij} \in \{0,1\}, \tag{9.3}$$

where

$$a_{ij} = \begin{cases} 1 & \text{if the UAV can fly over edge } (i,j). \\ 0 & \text{Otherwise} \end{cases}$$

The constraint in (9.1) indicates that at least one of the possible flying segments leaving the base (edge (0,*j*) such that $a_{0j} = 1$) is actually flown. The constraints in (9.2) guarantee a continuous route; if one of the incoming edges into a node *i* is used by the UAV, then one of the outgoing edges from *i* must be used too.

The additional constraints depend on the orientation of the model. In the time-oriented model the requirement is to maintain a minimum level α of survival probability. That is,

$$\sum_{i=0}^{W}\sum_{j=1}^{W+1}[\ln(1-p_{ij})]X_{ij} \geq \ln\alpha \,. \tag{9.4}$$

In the survivability-oriented model we set a time limit T_0 on the duration of the travel:

$$\sum_{i=0}^{W}\sum_{j=1}^{W+1} t_{ij}X_{ij} \leq T_0 \,. \tag{9.5}$$

Thus, the time-oriented model is Min $\sum_{i=0}^{W}\sum_{j=1}^{W+1} t_{ij}X_{ij}$ subject to (9.1), (9.2), (9.3), and (9.4), while the survivability-oriented model is Max $\sum_{i=0}^{W}\sum_{j=1}^{W+1}[\ln(1-p_{ij})]X_{ij}$ subject to (9.1) – (9.3), (9.5).

Example 1: A UAV is tasked to fly from its base to a given destination, using any of nine possible waypoints, as shown in Figure 1. The flying time and interception probability on each edge are given in Table 1.

Suppose that the mission is time critical and the UAV must reach the destination within 6 min. Solving the survivability-oriented problem, with $T_0 = 6$, we obtain the optimal route shown in Figure 2, with survivability probability of 0.32. However, if the mission is not time critical, but the required minimal mission-completion probability is 0.7, then we solve the time-oriented problem. Here the survivability probability is a constraint, completion time is the objective, and the optimal route is the one shown in Figure 3.

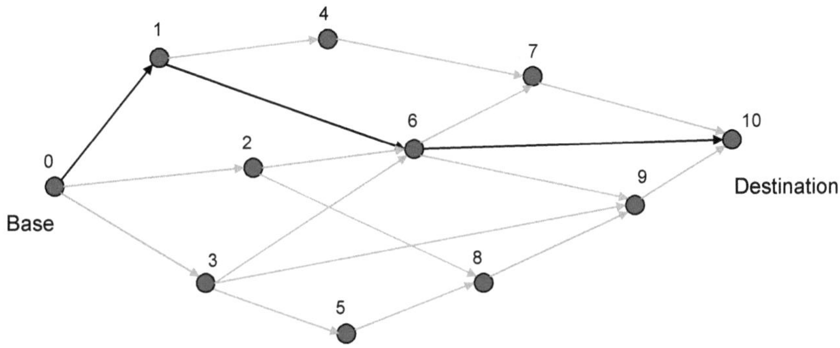

Figure 2: Optimal route, survivability-oriented model, $T_0 = 6$; flying time = 6 min; Pr[Survival] = 0.32.

Edge	Flying Time	Interception Probability
0,1	2	0.1
0,2	1	0.2
0,3	3	0.1
1,4	3	0
1,6	2	0.3
2,6	3	0.3
2,8	4	0.1
3,5	1	0.2
3,6	2	0.2
3,9	5	0.1
4,7	3	0.1
5,8	2	0.2
6,7	2	0.1
6,9	2	0.1
6,10	2	0.5
7,10	2	0.1
8,9	3	0.1
9,10	1	0.3

Table 1: Flying data.

The completion time for this route is 10 min, and the survival probability is 0.73, slightly more than the minimum threshold of 0.7. This survival probability (0.73) is also the highest possible over all routes.

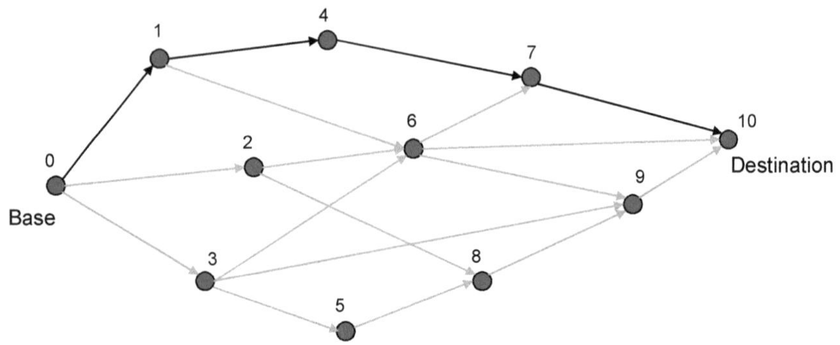

Figure 3: Optimal route, time-oriented model, $\alpha = 0.7$; flying time = 10 min; Pr[Survival] = 0.73.

Figure 4 presents the highest possible survivability probability for a given flying time, as obtained from solving a series of survivability-oriented problems. Notice the significant drop in this probability when the flying time changes from 10 min to 9 min

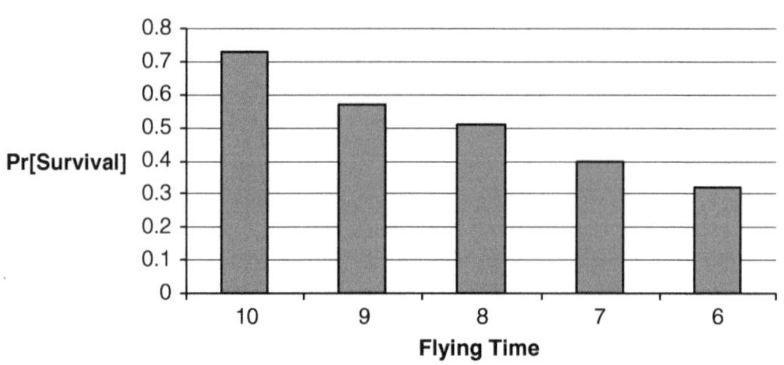

Figure 4: Maximum possible survival probability for a given flying time.

As was mentioned above, the maximum possible survivability probability, for any length of flying time, is 0.73. Flying time shorter than 6 min is infeasible. A chart of the type shown in Figure 4 can aid field commanders in deciding their best courses of action in routing a UAV for an intelligence mission. It demonstrates the trade-off between effectiveness (mission-completion time) and risk. Sheets "Intelligence-Time" and "Intelligence-Survivability" in *Chapter9.xls* solve the time-oriented and survivability-oriented problems, respectively. The models are solved using Excel's Solver.

9.2.2 Reconnaissance – Detecting a Mobile Target

Consider a team of Special Operations Forces (SOF) deployed in a remote desolate border area attempting to capture a suspicious person (e.g., a terrorist), henceforth called the *target*. According to intelligence reports, based on communication interception and human intelligence, the target is about to enter the area, attempting to cross it and infiltrate into the populated zone, with the objective of carrying out hostile action. The team is supported by a small UAV, which is launched immediately upon receiving the intelligence report. The UAV flies over the area and transmits live video pictures of its sensor's field of view to a ground control unit (GCU). The UAV is launched from the GCU.

The Problem. The mission of the UAV is to detect the target before it leaves the border area and enters the populated zone. Because the area is essentially deserted, and the UAV has a very small signature, we assume that it is not subject to interception. For the same reason we assume that the probability of a false-positive detection is negligible. In order to be able to transmit the video, the UAV must maintain line of sight with its GCU continuously, and it must fly within a certain range from that unit. The endurance of the UAV is limited, but we assume that it is longer than the time it would take the target to cross the area and disappear inland. The intelligence information regarding the target includes his velocity and a set of feasible routes. While the target's average velocity can be reasonably estimated based on his mode of travel (e.g., on foot, riding a horse, driving a vehicle), the specific route he may take is unknown. The objective of the SOF team is to plan a search pattern for the UAV that maximizes the probability of detecting the moving target. The target is assumed to be aware that he may be sought by a UAV and will choose his route to maximally frustrate the UAV team.

Figure 5 presents an example of possible entry points into the border area, and routes that the target may take in it.

There are three possible entry points into the border area – E1, E2, and E3. Each entry point is associated with one or two possible routes the target may take. For example, if the target enters the area through E1, then the two possible routes are (E1,a,D1) and (E1,a,c,D2). Entry point E2 has two associated routes and E3 has one route. Thus, there are altogether 2 + 2 + 1 = 5 possible routes in Figure 5. In addition to three entry points (E1, E2, E3) and three departure points (D1, D2, D3), there are four intermediate points – a,b,c,d.

The routes form a graph in which *nodes* are the entry points, departure points and intermediate points (so there are 10 nodes in Figure 5), and the *edges* are route segments connecting two adjacent nodes. The small circles labeled 1–4 are possible locations for the GCU. Let $r = 1,\ldots,R$ denote a route in the graph ($R = 5$ in Figure 5) and i, $i = 1,\ldots,I$ denotes an edge ($I = 9$ in Figure 5). Each edge belongs to one or more routes. Let l denote a feasible deployment location for the GCU, $l = 1,\ldots,L$ ($L = 4$ in Figure 5).

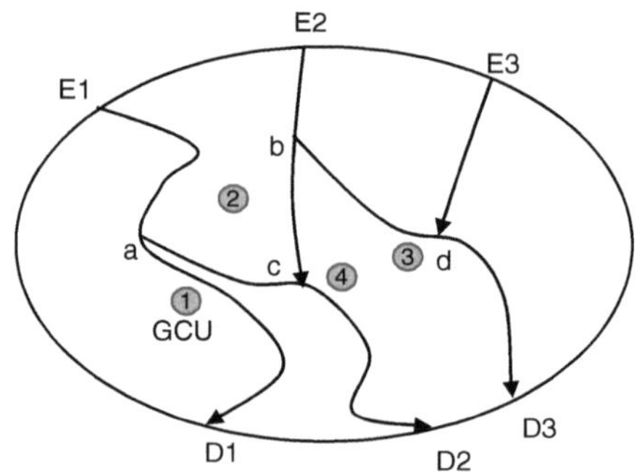

Figure 5: **Entry points, routes and possible GCU locations in the border area.**

We assume that the velocity of the target on each edge is known. Therefore, once the information that the target has entered the border area is received, and if his route is known, the SOF team can project exactly his whereabouts at any point in time (see Equation 9.6 below). The problem is that the target's route is unknown. We assume that time is discrete and that the SOF operation proceeds in time steps $t = 0,1,\ldots,T_0$, where T_0 is the time horizon of the operation (i.e., the maximum possible time the target spends in the border area).

The Model. Let,

$$c_{it}^{r} = \begin{cases} 1 & \text{if the target is on edge } i \text{ at time step } t\text{, given route } r \\ 0 & \text{Otherwise.} \end{cases} \tag{9.6}$$

The three-dimensional array $\{c_{it}^{r}\}$ is input data for the mission plan, which is obtained from analyzing the terrain and the distances of the various possible routes.

Consider now the UAV. We assume that in two consecutive time-steps the UAV can either stay in its current edge or move to an adjacent one. This assumption is not restrictive; the set of possible edges to which the UAV can move in one time step can be expanded if needed. For a given edge *i*, let J_i denote the set of edges that contains edge *i* and all edges to which the UAV can fly in the next time period. According to our assumption, this is the set of edge *i* and all of its outgoing edges. For example,

$$J_{(E2,b)} = \{(E2,b),(b,c),(b,d)\}\,.$$

Let D_i denote the set of GCU locations such that a UAV launched from such a location can reach edge *i* at the first time step, and let A_i denote the set of GCU locations that can maintain continuous communication with a UAV flying over edge *i*.

The probability that the UAV detects the target during a single time step, given the UAV and the target are present in edge *i* at the same time step, is q_i. The probabilities q_i, $i = 1,\ldots,I$ depend on the UAV sensor's field of view, its resolution, background clutter, etc. The detections between time steps are assumed to be independent. For example, if the UAV and the target were on edge i during T_i time steps and on edge j during T_j time steps, then the probability of detection would be $1-(1-q_i)^{T_i}(1-q_j)^{T_j}$.

Variables

The variables in this problem are binary:

$$X_{it} = \begin{cases} 1 & \text{if the UAV loiters over edge } i \text{ during time-step } t \\ 0 & \text{Otherwise} \end{cases}.$$

$$Z_l = \begin{cases} 1 & \text{if the GCU is deployed in location } l \\ 0 & \text{Otherwise} \end{cases}.$$

Objective Function

A natural MOE for this problem is the probability that the target is detected. However, to maximize this MOE we need to know the probability distribution of the routes the target may take. These data may not be available, and if it is, the problem becomes a nonlinear optimization problem, as shown later on. Instead, we consider an MOE that hedges against a worse-case detection scenario.

Given that the target has selected route r, the probability that he is *not* detected before leaving the search area is $\prod_{t=1}^{T_0}\prod_{i=1}^{I}(1-q_i)^{c_{it}^r X_{it}}$. The goal is to find a UAV search, and GCU deployment plan such that this probability, under a worse-case scenario, is minimized. Thus, the objective function is

$$\underset{X_{it},Z_l}{Min}\ \underset{r}{Max}\prod_{t=1}^{T_0}\prod_{i=1}^{I}(1-q_i)^{c_{it}^r X_{it}}. \tag{9.7}$$

This objective is quite conservative, since it in effect assumes that the infiltrator chooses his route to maximize his survival probability and does so in full knowledge of the UAV search plan. An alternative would be to treat this problem as a TPZS game (see Chapter 7) where the target's route and the searcher's search plan are each unknown when the other is selected. In terms of that game, we are calculating an upper bound on its value. As mentioned above, if there is some prior information about the likelihood of route r in the form of a probability α_r, $\sum_{r=1}^{R}\alpha_r = 1$, then another alternative would be to minimize the overall probability of no detection, that is,

$$\sum_{r=1}^{R} \alpha_r \prod_{t=1}^{T_0} \prod_{i=1}^{I} (1-q_i)^{c_{it}^r X_{it}}. \tag{9.8}$$

While the nonlinear objective function in (9.7) can be transformed into a linear one, the objective function in (9.8) cannot. Letting Y be the logarithm of the objective function, (9.7) is equivalent to

$$\begin{aligned} &\underset{X_{it}, Z_l}{Min}\, Y \\ &st \\ &\sum_{t=1}^{T_0} \sum_{i=1}^{I} \ln(1-q_i) c_{it}^r X_{it} - Y \le 0, \qquad r = 1,\dots,R. \end{aligned} \tag{9.9}$$

Since $\ln(1-q_i) < 0$, we can write (9.9) in an equivalent form

$$\begin{aligned} &\underset{X_{it}, Z_l}{Max}\, W \\ &st \\ &-\sum_{t=1}^{T_0} \sum_{i=1}^{I} \ln(1-q_i) c_{it}^r X_{it} - W \ge 0, \qquad r = 1,\dots,R. \\ &W \ge 0 \end{aligned} \tag{9.10}$$

Constraints

$$X_{i1} - \sum_{l \in D_i} Z_l \le 0, \ i = 1,\dots,I, \tag{9.11}$$

$$X_{it} - \sum_{j \in J_i} X_{j,t+1} \le 0, \ t = 1,\dots,T_0 - 1, \ i = 1,\dots,I, \tag{9.12}$$

$$\sum_{i=1}^{I} X_{it} \le 1, \quad t = 1,\dots,T_0, \tag{9.13}$$

$$M \sum_{l \in A_i} Z_l - \sum_{t=1}^{T_0} X_{it} \ge 0, \ i = 1,\dots,I, \tag{9.14}$$

$$\sum_{l=1}^{L} Z_l \le 1. \tag{9.15}$$

Constraints (9.11) and (9.12) guarantee a feasible search plan for the UAV. From constraints (9.11) it follows that the UAV can search an edge during the first time step only if the edge is reachable on time from the launching location. Similarly to constraints (9.2) in Section 9.2.1 constraints (9.12) guarantee a feasible continuous search pattern. The UAV can search edge *i* during time step *t*, ($X_{it} = 1$), only if it searches that edge, or one of its adjacent edges, at time $t-1$. Constraints (9.13) limit the presence of the UAV to at most one edge at any given time step. Constraints (9.14) restrict the searched edges only to those which are within range and line of sight to the GCU. *M* is a large constant. Constraint (9.15) assures that the GCU is deployed in at most one location.

Thus the optimization problem is (9.10) – (9.15), with $X_{it}, Z_l \in \{0,1\}$, which is a linear mixed-integer (0-1) programming problem.

Example 2: Consider the scenario shown in Figure 5. Table 2 presents base case data. For each edge *i* the table presents the target's travel time, the set of GCU locations from which a UAV can reach edge *i* during the first time step, D_i, and the GCU locations that can maintain continuous communication with a UAV flying over edge *i*, A_i.

Edge	(E1,a)	(a,D1)	(a,c)	(E2,b)	(b,c)	(c,D2)	(b,d)	(E3,d)	(d,D3)
Travel time	3	4	3	2	4	4	4	5	3
D_i	1,2,4	1,2,4	1,2,3,4	2,3,4	1,2,3,4	1,2,3,4	2,3,4	3,4	1,3,4
A_i	1,2,3,4	1,2,3,4	1,2,3,4	2,3,4	1,2,3,4	1,2,3,4	1,2,3,4	2,3,4	1,2,3,4

Table 2: **Travel time and accessibility of edges – base case.**

It is easily verified that the time horizon is $T_0 = 10$. We assume that the (instantaneous) probability of detection on each edge is the same, $q = 0.7$. Solving the optimization problem (9.10) – (9.15) we obtain the following UAV optimal search pattern for the base case (Table 3):

Edge	(E1,a)	(a,D1)	(a,c)	(E2,b)	(b,c)	(c,D2)	(b,d)	(E3,d)	(d,D3)
Time step	1,2	-	3	-	4,5	-	6	-	7,8,9,10

Table 3: **Optimal search pattern for the base case**

The GCU is deployed in location 1. Once launched from location 2, the UAV searches edge (E1,a) during the first two time steps. Then the UAV flies over (a,c) for one time period, (b,c) for two time periods, and (b,d) for one time period. The last four time steps are spent in (d,D3). The probability of detecting the target in a worst-case scenario is 0.91, which means that at each one of the five possible routes the UAV has at least two detection opportunities if the target selects that route.

Suppose now that the topography at the border area is more rugged and the communication possibility between the GCU locations and the various edges is more restricted, as shown in Table 4. The new updated location for the GCU is 3. The new optimal search pattern is also shown in Table 4. The detection probability in the worst-case scenario is now 0.7, which means that in at least one scenario (route) there is only one detection opportunity.

Edge	(E1,a)	(a,D1)	(a,c)	(E2,b)	(b,c)	(c,D2)	(b,d)	(E3,d)	(d,D3)
A_i	2,3	1,2,3	1,2,3,4	2,3,4	1,3,4	1,2,4	1,2,3,4	2,3,4	1,2,3,4
Time step	2,3	-	1,4	-	5	-	6	-	7,8,9,10

Table 4: Optimal search pattern for restricted flying zones.

9.3 Unmanned Combat Aerial Vehicles

An unmanned combat aerial vehicle (UCAV) is a self-propelled aerial vehicle that typically loiters over the target area, seeking targets. In addition to the functionality of a UAV, a UCAV is also a weapon, which can attack targets on the ground or at sea. A UCAV may be *retrievable* or *disposable*. Retrievable UCAVs are relatively large UAVs that carry one or more munitions such as bombs or missiles. They are launched toward the target area in a controlled trajectory, and upon arrival in the target area, they start their search for targets to attack with the munitions they carry. Once their munitions are expended, the UCAVs return to their base for refit and reload. Disposable UCAVs are essentially precision-guided munitions, where the warhead is an integral part of the platform. Clearly, a disposable UCAV can attack at most one target. In this section we describe two UCAV models: (1) disposable vehicle and (2) retrievable multiple-weapon vehicle.

In the following models we assume that there are two types of targets in the target area: *valuable targets* (VT) and *non-valuable targets* (NVT). The definition of a VT is derived from the scenario and the mission objectives. For example, in some situations armored vehicles would be considered as VTs, while in other situations surface-to-air missile batteries would be VTs. All other objects in the target area that may be considered by the UCAV as targets are NVTs. In particular, a killed VT becomes NVT. The situation we consider is where a UCAV loiters over the target area searching for VTs. Once the UCAV detects an object, it investigates the object and determines if it is a VT or an NVT. According to the classification of the target, the UCAV engages the object or abandons it and continues with the search. If the UCAV engages an object, it attacks it and kills it with a certain probability. The sensor of the UCAV is imperfect. It may erroneously identify a VT as an NVT (false-negative error) and an NVT as a VT (false-positive error).

9.3.1 Disposable UCAV

A single disposable UCAV is launched to attack valuable targets (VTs) in a certain area. The area contains also non-valuable targets (NVTs). We initially assume that the UCAV is memoryless; it does not keep track of earlier non-engagement decisions and therefore may revisit and reinvestigate previously detected objects.

The Problem. We wish to compute the probability that the UCAV successfully attacks a VT within a certain given time window. The inputs of the problem are (1) number of VTs and NVTs in the target area, (2) UCAV's mean inter-detection time – the time it takes the UCAV to move from one target to another, (3) UCAV's kill probability, and (4) false-negative and false-positive detection probabilities.

The Model. There are T targets in the target area; out of which K targets are VTs and $T-K$ are NVTs. The UCAV loiters over the target area and detects targets at a rate λ. Following detection, the UCAV instantly identifies the object and decides whether to engage it or abandon it and move on to the next target. The inter-detection time is exponentially distributed with mean $1/\lambda$, which implies that the number of detections in a certain time interval Δt is a Poisson random variable with parameter $\lambda\Delta t$. The probability of correct identification by the UCAV is q and r for a VT and NVT, respectively. The parameter q represents the *sensitivity* of the UCAV's sensor ($1-q$ is the false-negative probability), while r represents its *specificity* ($1-r$ is the false-positive probability). In other words, if the UCAV detects a VT, it engages it with probability q, and if it detects an NVT, it engages it (erroneously) with probability $1-r$. Recall that the UCAV is memoryless; it may revisit previously detected (and identified as NVT) objects, and possibly engage them. The kill probability of a VT, given the target has been engaged, is p. For a mission time window of duration τ, we compute the probability that the UCAV kills a VT during that time window.

Three possible outcomes may occur following a detection event:

- Correct engagement of a VT, with probability: $\varphi_1 = \frac{K}{T}q$;
- Erroneous engagement of an NVT, with probability: $\varphi_2 = \frac{T-K}{T}(1-r)$;
- No engagement, with probability: $\varphi_3 = \frac{K}{T}(1-q) + \frac{T-K}{T}r = 1-\varphi_1-\varphi_2$.

Note that these probabilities depend on the ratio between the number of VTs and NVTs and not on their absolute values.

Given n detection opportunities during the time window, the probability that the UCAV engages a VT is

$$P_n[VT] = \sum_{i=1}^{n} \varphi_3^{i-1}\varphi_1 = \frac{1-\varphi_3^n}{1-\varphi_3}\varphi_1. \quad (9.16)$$

The *ith* term in the sum in (9.16) is the probability that the *ith* detection will result in correctly engaging a VT, which requires that the first $i-1$ detections result in no engagement. The second equality in (9.16) is because the sum is a geometric series. When $n \to \infty$ (i.e., when there is no time constraint for the mission and the UCAV has unlimited endurance) then $P_n[VT] \to \frac{\varphi_1}{1-\varphi_3} = \frac{\rho q}{1+\rho q - r}$, where $\rho = \frac{K}{T-K}$ is the ratio between the number of VTs and NVTs in the target area.

Recall that the number of detections follows a Poisson distribution, therefore, given a time window τ, absent engagement, the probability of n detections is

$$\Pr[N = n] = \frac{(\lambda\tau)^n e^{-\lambda\tau}}{n!}. \tag{9.17}$$

From (9.16) and (9.17) we obtain that the (unconditional) probability that the UCAV engages a VT is

$$\begin{aligned} P_\tau(VT) &= \sum_{n=0}^{\infty} \frac{1-\varphi_3^n}{1-\varphi_3}\varphi_1 \frac{(\lambda\tau)^n e^{-\lambda\tau}}{n!} = \frac{\varphi_1}{1-\varphi_3}\left(1-e^{-\lambda\tau(1-\varphi_3)}\right) = \\ &= \frac{\rho q}{1+\rho q - r}\left(1-e^{-\lambda\tau\left(1-\frac{\rho}{1+\rho}(1-q)-\frac{1}{1+\rho}r\right)}\right) \end{aligned} \tag{9.18}$$

and the probability a VT is killed is $pP_\tau(VT)$. From (9.18) we can also observe that the time until an object is engaged is exponentially distributed random variable with mean $1/(\lambda(1-\varphi_3))$.

Example 3: The ratio between the number of VTs and NVTs in the target area is $\rho = 1$. The detection rate is $\lambda = 1$ per minute, and the probabilities are q = 0.7, r = 0.8, and p = 0.5. The time window is $\tau = 10$ min. The kill probability is $pP_{10}(VT) = 0.5 \times 0.77 = 0.385$. (See sheet "Disposable Memoryless" in *Chapter9.xls*). Figure 6 presents the effect of the UCAV's sensitivity (q) and specificity (r) on the probability of engaging a VT.

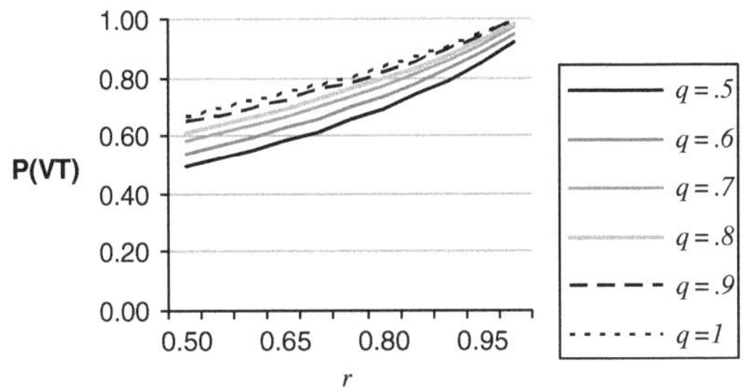

Figure 6: The probability of engaging a VT as a function of *r* and *q*.

Notice that the effect of the two error probabilities is not symmetrical. When $r = 0.5$ and $q = 1$, $P_{10}(VT) = 0.67$, while in the opposite direction, $r = 1$ and $q = 0.5$, we obtain $P_{10}(VT) = 0.92$. From Figure 6 we can conclude that specificity has a larger effect than sensitivity on the performance of the UCAV. This comes with no surprise since false-negative error can be rectified later on in the mission, while false-positive error cannot; engaging the wrong target (NVT) by the UCAV results in a mission failure.

Figure 7 shows the effect of the target ratio ρ on the VT engagement probability, for three values of detection rate λ.

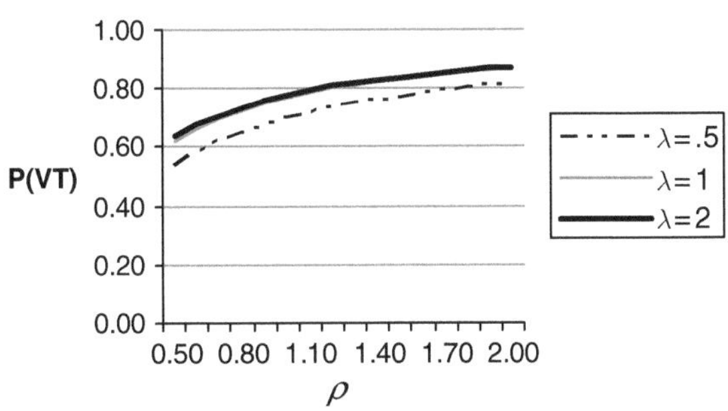

Figure 7: The probability of engaging a VT as a Function of ρ and λ.

We can see from Figure 7 that detection rate higher than 1 does not affect the VT engagement probability; this probability converges very rapidly to $\dfrac{\varphi_1}{1-\varphi_3}$.

Suppose now that the UCAV has *perfect* memory and it would never revisit a target that has been detected before and has been identified as NVT. The UCAV selects a target at random from the set of targets not detected yet. The maximum possible number of detections in this case is T, and we assume that the time window for the mission is such that this maximum can be attained. Note that the T detections may result in no engagement, if all K VTs are misclassified by the UCAV as NVTs, and all $T-K$ NVTs are identified correctly. In that event, which occurs with probability $(1-q)^K r^{T-K}$, we assume that the UCAV selects at random a target and engages it. Unlike the memoryless case, where the probability of engaging a VT only depends on the ratio between the number of VTs and NVTs, here this probability depends on the absolute numbers of VTs and NVTs, that is, $P(VT)=P(K,T)$. Let $Q(K,T)$ be the probability of a successful engagement before the search is over. This probability is obtained recursively:

$$Q(K,T)=\frac{K}{T}(q+(1-q)Q(K-1,T-1))+\frac{T-K}{T}rQ(K,T-1). \tag{9.19}$$

In (9.19), the boundary conditions are that $Q(0,t)=0,\ t=0,\dots,T$ and $Q(1,1)=q$. The first term on the right-hand side of (9.19) represents the situation where a VT is detected but is erroneously identified as an NVT. The second term represents the situation where an NVT is detected and is correctly identified as such. In both cases the search for a VT continues. The probability $P(K,T)$ of engaging a VT is given now by,

$$P(K,T)=Q(K,T)+\frac{K}{T}(1-q)^K r^{T-K}. \tag{9.20}$$

Example 4: There are T = 10 targets in the target area, out of which K = 5 are VTs. The sensitivity and specificity probabilities are $q=r=0.75$. The probability of engaging a VT is 0.769 (See sheet "Disposable Full Memory" in *Chapter9.xls*). From (9.16) it follows that the maximum possible engagement probability (when $n\to\infty$) for the memoryless UCAV is 0.75. Figure 8 presents the values of $P(VT)$ for the situation where the number of VTs is equal to the number of NVTs, and the total number of objects range from 2 to 20. Clearly, the full memory UCAV outperforms the memoryless one but asymptotically approaches its value –0.75. Notice also that the full memory UCAV performs best in this case when the number of objects is 4.

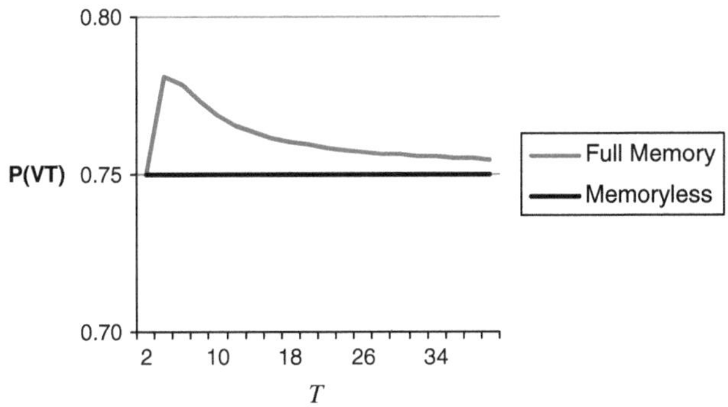

Figure 8: The probability of engaging a VT for the two types of UCAVs, $q = r = 0.75$.

Figures 9 and 10 depict these probabilities for the situations $q = 1$, $r = 0.5$, and $q = 0.5$, $r = 1$, respectively.

We observe that the situation of perfect specificity ($r = 1$) and poor sensitivity ($q = 0.5$) totally dominates the opposite situation of poor specificity ($r = 0.5$) and perfect sensitivity ($q = 1$). This observation leads to the conclusion that it is much more effective to invest in reducing the false-positive error than investing in reducing the false-negative error. Moreover, while in the first case (Figure 9), the full memory UCAV dominates the memoryless one, as in the base case above, the reverse is true in the second case (Figure 10) when the specificity is perfect. This is no surprise; if the UCAV can always detect an NVT, eventually it will acquire a VT.

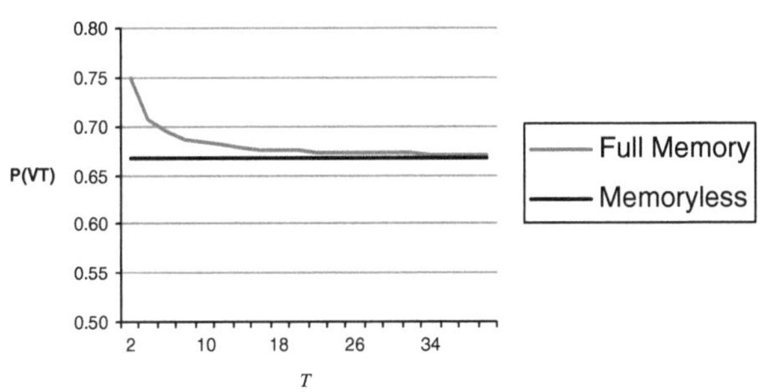

Figure 9: The probability of engaging a VT for the two types of UCAVs, $q = 1$, $r = 0.5$.

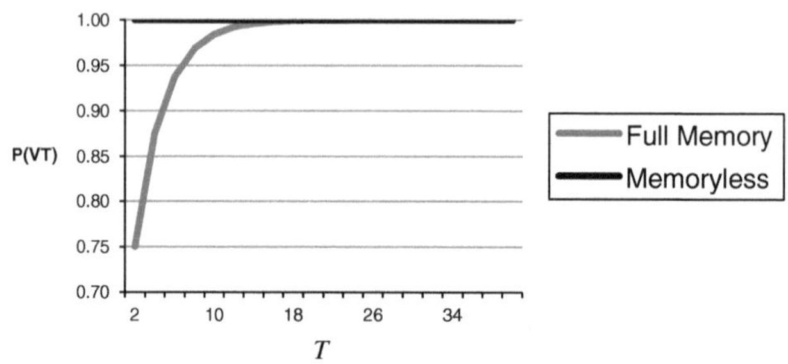

Figure 10: The probability of engaging a VT for the two types of UCAVs, $q = 0.5$, $r = 1$.

9.3.2 Retrievable Multiple-Weapon UCAV

A retrievable UCAV carrying a number of munitions loiters over the target area in search of VTs. Once the UCAV detects a target and identifies it as a VT, the UCAV engages it with one of its munitions. The UCAV is memoryless and detects the targets in a random (uniform) manner.

The Problem. Given a certain number of VTs and NVTs in the target area, and a number of munitions on-board the UCAV, the objective is to compute the probability distribution function of the number of killed VTs, and the expected duration of the operation. Assuming a long endurance of the UCAV relative to the duration of a typical mission, the operation ends when the last munition is delivered.

The Model. Let T denote the total number of objects (VTs and NVTs) in the target area, and let K denote the initial number of VTs. Thus, $T-K$ is the initial number of NVTs. The UCAV carries N munitions at the beginning of the operation, and the probability of correct identification by the UCAV is q and r for a VT and NVT, respectively. The mean inter-detection time – the time between two consecutive detections – is E_D. The mean attack time – the time from detection until the munition is delivered – is E_A. The kill probability is p.

We develop now a model for obtaining the probability distribution of the number of killed VTs (in particular, its expected value) and the expected duration of the operation. Suppose that at a certain stage in the engagement there are $k, k \leq K$, VTs, and $T-k$ NVTs. The probability that the UCAV engages a VT given detection is $\frac{k}{T}q$, and the probability of engaging an NVT is $\frac{T-k}{T}(1-r)$.

Next define a Markov chain (see Appendix A) that describes the UCAV's operation. A *step* in this chain is a detection event, and a *state* is a pair (n,k), where n

is the number of munitions left on-board the UCAV and k is the number of VTs left in the target area (recall that a killed VT is NVT). There are three possible transitions following a detection event.

$$(n,k) \rightarrow \underbrace{(n-1,k-1)}_{\text{The UCAV killed a VT}} \text{ with probability } \frac{k}{T}qp, \tag{9.21}$$

$$(n,k) \rightarrow \underbrace{(n-1,k)}_{\substack{\text{The UCAV acquired} \\ \text{a VT but missed it,} \\ \text{or acquired an NVT}}} \text{ with probability } \frac{k}{T}q(1-p)+\frac{T-k}{T}(1-r), \tag{9.22}$$

$$(n,k) \rightarrow \underbrace{(n,k)}_{\substack{\text{The UCAV did not} \\ \text{acquire the target}}} \text{ with probability } \frac{k}{T}(1-q)+\frac{T-k}{T}r. \tag{9.23}$$

The initial state is (N, K) and the absorbing states (see Appendix A) are $(0, k)$, $k = Max\{0, K-N\},\ldots,K$. Note that if the UCAV has perfect sensitivity and specificity, that is, $q = r = 1$, $N > K$, and the number K of VTs is known to the UCAV, then the states $(n,0), n = 1,\ldots,N-K$ are absorbing too since the UCAV will leave the target area once all the VTs are killed. The possible states are

$$\begin{array}{l} (N,K) \\ (N-1,K),(N-1,K-1) \\ (N-2,K),(N-2,K-1),(N-2,K-2). \\ . \\ . \\ (0,K),(0,K-1), \quad \ldots \quad ,(0,Max\{0,K-N\}), \end{array} \tag{9.24}$$

and the number of states is

$$S = \begin{cases} (K+1)\left(\dfrac{K+2}{2}+N-K\right), & \text{if } N > K \\ \dbinom{N+2}{2}, & \text{if } N \le K \end{cases}. \tag{9.25}$$

From now on we assume that the number of munitions loaded on the UCAV is not larger than the initial number of VTs in the target area, that is, $N \le K$. In that case, the number of absorbing states is $N+1$ (see the last row of (9.24)), and the number of transient states (see Appendix A) is $\dbinom{N+1}{2}$.

The Markov transition matrix M can be written as

$$M = \begin{bmatrix} I_{N+1} & 0 \\ R & Q \end{bmatrix}, \tag{9.26}$$

where I_{N+1} is a $(N+1)\times(N+1)$ unit matrix that corresponds to the absorbing states, Q is a $\binom{N+1}{2}\times\binom{N+1}{2}$ matrix corresponding to the transient states, and R is a $\binom{N+1}{2}\times(N+1)$ matrix that represent transitions from a transient state to an absorbing one. The probability distribution of the number of VTs killed is given in the first row of $(I-Q)^{-1}R$, where I is a $\binom{N+1}{2}\times\binom{N+1}{2}$ unit matrix, see Appendix A. Note that the first row of $(I-Q)^{-1}R$ corresponds to the initial state of the mission before the first munition is delivered. The expected number of visits to a transient state (*n*,*k*), $\mu_{(n,k)}$ is the (*n*,*k*)-th entry in the first row of $(I-Q)^{-1}$ (see Appendix A), therefore, the expected duration of the mission is

$$E[Time] = E_D \sum_{(n,k)\, transient} \mu_{(n,k)} + NE_A. \tag{9.27}$$

Example 5: A UCAV carrying $N = 3$ munitions is sent out to attack a target area, which comprises $T = 12$ targets, out of which $K = 6$ are VTs. The probability of recognizing a VT is $q = 0.7$, and the probability of recognizing an NVT is $r = 0.8$. The kill probability is $p = 0.5$. The mean inter-detection time is $E_D = 3$ min, and the mean attack time is $E_A = 0.5$ min. According to (9.25) there are 10 states, out of which 4 states are absorbing. The transition matrix (see sheet "Retrievable" in *Chapter9.xls*) is

$M =$

States	(0,6)	(0,5)	(0,4)	(0,3)	(3,6)	(2,6)	(2,5)	(1,6)	(1,5)	(1,4)
(0,6)	1	0	0	0	0	0	0	0	0	0
(0,5)	0	1	0	0	0	0	0	0	0	0
(0,4)	0	0	1	0	0	0	0	0	0	0
(0,3)	0	0	0	1	0	0	0	0	0	0
(3,6)	0	0	0	0	0.55	0.28	0.18	0	0	0
(2,6)	0	0	0	0	0	0.55	0	0.28	0.18	0
(2,5)	0	0	0	0	0	0	0.59	0	0.26	0.15
(1,6)	0.28	0.18	0	0	0	0	0	0.55	0	0
(1,5)	0	0.26	0.15	0	0	0	0	0	0.59	0
(1,4)	0	0	0.25	0.12	0	0	0	0	0	0.63

The highlighted area in M is the sub matrix Q. The matrix $I-Q$ is

$$\begin{matrix} 0.45 & -0.3 & -0.2 & 0 & 0 & 0 \\ 0 & 0.45 & 0 & -0.3 & -0.18 & 0 \\ 0 & 0 & 0.41 & 0 & -0.26 & -0.15 \\ 0 & 0 & 0 & 0.45 & 0 & 0 \\ 0 & 0 & 0 & 0 & 0.41 & 0 \\ 0 & 0 & 0 & 0 & 0 & 0.37 \end{matrix}$$

and its inverse $(I-Q)^{-1}$ is

$$\begin{matrix} 2.22 & 1.36 & 0.95 & 0.83 & 1.19 & 0.38 \\ 0 & 2.22 & 0 & 1.36 & 0.95 & 0 \\ 0 & 0 & 2.45 & 0 & 1.57 & 0.97 \\ 0 & 0 & 0 & 2.22 & 0 & 0 \\ 0 & 0 & 0 & 0 & 2.45 & 0 \\ 0 & 0 & 0 & 0 & 0 & 2.73 \end{matrix}$$

Using (9.27) we obtain that the expected duration of the operation is

$$E[Time] = 3\times(2.22+1.36+0.95+0.83+1.19+0.38)+3\times0.5 = 22.3 \text{ min}$$

The probability distribution is obtained from the first row of $(I-Q)^{-1}R$:

States (0,6)	(0,5)	(0,4)	(0,3)
0.23	0.46	0.27	0.04
0.37	0.49	0.14	0.00
0.00	0.41	0.47	0.11
0.61	0.39	0.00	0.00
0.00	0.64	0.36	0.00
0.00	0.00	0.68	0.32

The probability that each one of the three UCAVs kills a VT (state (0,3)) is 0.04, while the probability that none is successful (state (0,6)) is 0.23.

9.4 Summary, Extensions and Further Reading

In this chapter we presented two types of UAV models: prescriptive models for UAV routing in Section 9.2 and descriptive models for autonomous UCAV deployment in Section 9.3. These simple and basic models exhibit major factors that dominate UAV operations – time, survivability, detection capabilities, and memory – and highlight the main features of UAV modeling – routing, scheduling, and effectiveness assessment. These models can be expanded in several ways, described briefly below.

9.4.1 Optimizing the Employment of UAVs

While the models in Section 9.2 reflect realistically the objective functions of UAV operations such as time to complete a mission, probability of mission completion and the probability of detecting a target, they ignore many constraints that are typical in such settings. UAVs may have limited endurance due to logistical constraints, therefore refit and refuel actions must also be represented in an optimization model. There may also be physical constraints, which are associated with the specific flying capabilities of the UAV, not just the topography, that will limit its maneuverability and thus constrain its feasible routes. Also, the models presented are two dimensional in the sense that only the projection on the ground of the location of the UAV – the (i, j) coordinate – is taken into account. In reality the altitude of the UAV affects its capability to accomplish its mission and therefore must also be taken into account.

Moreover, the models in Section 9.2 optimize the employment of only one UAV. In many combat situations the ground force may operate more than one. The challenge in such situations is to efficiently coordinate the route and schedule of each UAV such that the objective function is optimized. In such multiple-UAV situations additional constraints relate to flying safety – deconfliction among the flying patterns of the various UAVs, control – assigning UAVs to ground control centers and, most of all, managing and fusing effectively the information obtained from the UAVs' sensors. Kress and Royset (2008) present a model that addresses some of the extensions mentioned above in a multiple-UAV setting. Alighanbari and How (2008) present a robust approach to task assignment of UAVs, and Rasmussen and Shima (2008) develop a tree search algorithm for assigning cooperating UAVs to multiple tasks. Optimal deployment and employment of UAVs for search missions is a special case of search problems addressed in Chapter 7. Dell et al. (1996) study multiple searches in constrained path moving-target settings that might be applied to UAVs.

9.4.2 Unmanned Combat Aerial Vehicles

In Section 9.3 we presented two descriptive models for a single UCAV. In reality UCAVs, in particular disposable UCAVs, are deployed in swarms that loiter over the target area. Similarly to multiple search UAVs, a key question is how to coordinate the UCAVs and what is the effect of such coordination on the outcome of the mission. A simple extension of the model in Section 9.3.1 is where a swarm of disposable UCAVs attack a cluster of targets (e.g., armored vehicles in a battalion). Similarly to the Markov model in Section 9.3.2, a state of the battle would be represented by the pair (n, k) where n is the number of UCAVs and k is the number of VTs. However, unlike the single retrievable UCAV with multiple munitions in Section 9.3.2, absent coordination, two or more disposable UCAVs may acquire simultaneously and independently the same VT and thus waste valuable attack resources. This situation leads to a more complex Markov model that is described in Kress et al. (2006b).

The models described in Section 9.3.2 and in the paper mentioned above are limited to homogeneous targets and homogeneous UCAVs. At significant computational cost, these models may be extended to account for non-homogeneous targets and multiple types of UCAVs. Another interesting and potentially important extension is to incorporate decision rules where the UCAVs manifest some level of cognitive capability. Specifically, in reality both sensitivity and specificity probabilities may depend on the time a UCAV spends investigating a target. This aspect is not captured in current models and may lead to interesting optimization models.

Exercises

(1) Consider Example 1 and suppose that due to operational or safety considerations the UAV cannot fly over edge (0,1). Use sheet "Intelligence-Survivability" of *Chapter9.xls* to answer the following questions (*Hint*: assign t_{01} a very large value).

(a) Can the mission be completed within 8 min with survivability probability not less than 0.7? Ans. No.

(b) If the mission is to be completed within 7 min, what is the optimal route and what is the probability the UAV reaches its destination unharmed? Ans. (0,2),(2,6),(6,9),(9,10); 0.353.

(c) What is the fastest route if the required probability for mission completion is 0.6? Ans. No route is feasible.

(2) Consider Example 1. Based on additional intelligence information, it is decided that the route of the UAV *must* pass through waypoint 8. Write an additional constraint for (9.1) – (9.3) that reflects this requirement.
Ans. Require that $X_{28} + X_{58} = 1$. The optimal solution should have the UAV flying from waypoint 8 to waypoint 9 ($X_{89} = 1$).

(3) The military is considering adopting one of two possible disposable UCAVs: UCAV1, which is memoryless, with limited endurance but with high specificity, and UCAV2 which has unlimited endurance, full memory but with lower specificity. Specifically, UCAV1's detection rate is $\lambda = 1$, its endurance is $\tau = 15$ min and its specificity is $r = 0.95$. The specificity of UCAV2 is $r = 0.7$. The sensitivity and kill probability are the same for both UCAVs, $q = 0.7$ and $p = 0.5$, respectively. All other parameters (e.g., cost, maintenance, interoperability) are assumed to be equal. The reference scenario is a target area with 10 targets out of which 5 are VTs. Which UCAV is more effective? Use the two "Disposable" sheets of *Chapter9.xls*.
Ans. UCAV1. $P_K(\text{UCAV1}) = 0.465, P_K(\text{UCAV2}) = 0.359$.

(4) Consider Example 5 with $T = 30$ targets and $K = 17$ VTs. (a) compute the expected number of killed VTs, (b) plot a graph of the expected number of killed VTs as a function of the number of VTs in the target area.
Ans. (a) 1.22.

Chapter 10
Terror and Insurgency

Fighting terrorism is like being a goalkeeper. You can make a hundred brilliant saves but the only shot that people remember is the one that gets past you.

Paul Wilkinson

10.1 Introduction

So far we have studied "pure" military situations in the sense that all actors are established state military entities such as batteries of surface-to-air missile launchers, infantry battalions, air force squadrons, swarms of UAVs or navy ships. The military engagements, combat situations, and conflicts described in the previous chapters were typically between such entities. In this chapter we broaden the scope and consider non-state violent actors such as terrorist groups, guerillas, and insurgents, collectively called in this section – *rebellions*.

Terrorist acts against the civilian population and insurgency activities against the state are committed by stateless individuals or organizations. Since the inhabitable parts of the Earth are partitioned into states, rebellions must necessarily reside in some "host" state. If the host state is itself the rebellion's target, then these stateless actors must be careful to disguise themselves from the state's police or armed forces and to rely on the population for support. It follows that rebellions' weapons must be relatively simple and small – they must maintain a small signature in order to thrive and operate. Rebellions do not maintain inventories of aircraft or tanks or ships, like many states do, but rely instead on small arms, improvised explosive devices, biological agents, poisons, and other methods of causing mayhem that do not require persistent exposure.

Countering rebellions by a regular state force is sometimes called *asymmetric* or *irregular warfare*. The asymmetry is manifested in three main dimensions: force size, military capabilities, and intelligence. In terms of physical net assessment, the capabilities of the rebels are no match to those of the state forces; the number of combatants available to the regime forces is typically at least an order of magnitude larger than the number of active rebels (as of July 2007 there were more than 500,000 Coalition and Iraqi Security Forces operating in Iraq, while the estimate for the number of insurgents ranged between 15,000 and 70,000, O'Hanlon and Campbell, 2007). The state forces are usually better equipped and trained, and they operate more advanced and effective weapon systems than the

A. Washburn, M. Kress, *Combat Modeling*, International Series in Operations Research & Management Science 134, DOI 10.1007/978-1-4419-0790-5_10,

rebellions. The only advantage of the rebellions is their elusiveness and invisibility; the regime forces have the military means and capabilities to effectively engage the targets, but they simply cannot find them. In other words, the regime forces lack situational awareness. The rebels, on the other hand, have many exposed targets to choose from and their situational awareness regarding their civilian and military targets is almost perfect.

In this chapter we address three typical problems associated with terrorism and insurgency. Section 10.2 presents a model that estimates the effect of a suicide bomber in a crowded area. In particular, the effect of *crowd blocking* is modeled and evaluated. The second problem concerns biological terrorism in the form of intentional spread of contagious agents. Section 10.3 discusses this problem in the context of smallpox and presents an epidemiological model that is based on the classical susceptible–infected–recovered (SIR) model (Anderson and May, 1991). The epidemiological model describes a mass vaccination process and evaluates the number of casualties, as well as the required isolation capacity, as a function of the vaccination capacity and other parameters. Finally, in Section 10.4 we present a simple dynamic, Lanchester-based, model that describes the evolution of an insurgency and of the attitude of the population in the presence of the regime's counterinsurgency operations.

10.2 The Effect of Suicide Bombing

Terror events that involve suicide bombers (SB) are a major concern to regimes – mostly because of their psychological effect on the population, which may be amplified by media coverage. Typical suicide bomb events occur in relatively small, crowded areas, henceforth called *arenas*, such as restaurants, buses, and bus stations. AnSB enters an arena and attempts to detonate the bomb he carries so as to maximize the number of casualties. In this section we will construct a model that is intended to explore the effect of such a bomb as a function of various relevant parameters, especially the crowd density in the arena.

One would expect that the number of casualties would increase with the density of the crowd, although at a decreasing rate, and would decrease with the size of the arena for a fixed number of people. The reason for the first assertion is that a higher density of people increases the probability that a random fragment of the bomb hits a person, and therefore it also increases the expected number of casualties. The second assertion seems reasonable for a similar reason; for a fixed crowd size, a larger arena implies lower density.

We will show that this is not necessarily the case. *Crowd blocking* has a significant effect on the expected number of casualties. Crowd blocking occurs when some persons are shielded from the fragments of the bomb by other persons who stand between them and the SB. The SB model presented here is reported in Kress (2005a). An extension of the model that examines the effectiveness of SB mitigation strategies is described in Kaplan and Kress (2005).

10.2.1 Dispersion of Fragments

A typical SB carries an explosive charge (EC) mounted on a belt, which is concealed under a coat or a large shirt. The EC contains explosive and small pieces of metal, such as screws and nails. The SB attempts to position himself as close to the center of the crowd as possible, and then to set off the bomb. The metal fragments in the bomb disperse in a beam spray that hit people in the vicinity of the SB. Due to the relatively small amount of explosive (3–4 kg) we assume that the energy of the fragments is such that a fragment may injure an exposed person or, with a lower probability, a person standing right behind an exposed person. The number of *effective fragments* – those which are potentially effective – is roughly one half the number of fragments on the belt. The other half is wasted on the SB himself. We assume that the explosive and fragments are uniformly distributed on the belt, which is depicted as a full circle; see Figure 1.

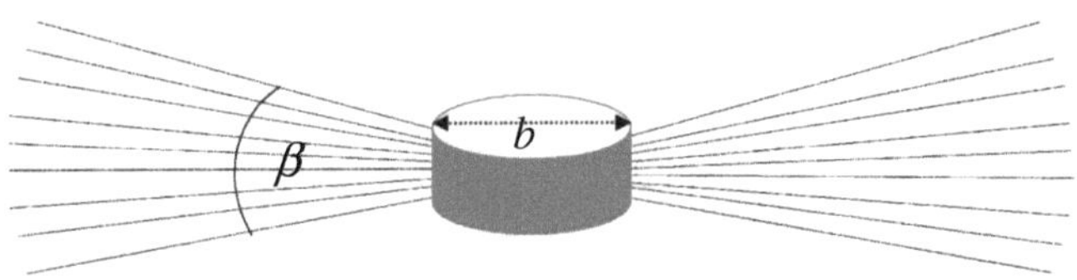

Figure 1: Suicide belt and its spray beam.

Let
N – Number of effective fragments in the beam spray
R – Range from the belt of the SB
b – Diameter of the suicide belt (the average "diameter" of a person's body)
β – Dispersion angle of the effective beam spray
$P_H(R)$ – Probability that an exposed person at range R is hit

First we compute the density of fragments at a range R from the SB.

As shown in Figure 1, the beam spray is distributed vertically with angle α between $-\frac{\beta}{2}$ and $+\frac{\beta}{2}$ from the horizontal at the point of explosion. This is the center of the sphere in Figure 2. The arc length at range R is $ds = Rd\alpha$. The surface area of a complete circumferential strip at angle α is $2\pi R\cos\alpha R d\alpha = 2\pi R^2 \cos\alpha d\alpha$. The total area of dispersion at range R is therefore

$$2\int_0^{\frac{\beta}{2}} 2\pi R^2 \cos\alpha d\alpha = 4\pi R^2 \int_0^{\frac{\beta}{2}} \cos\alpha d\alpha = 4\pi R^2 [\sin\alpha]_0^{\frac{\beta}{2}} = 4\pi R^2 \sin\frac{\beta}{2} \tag{10.1}$$

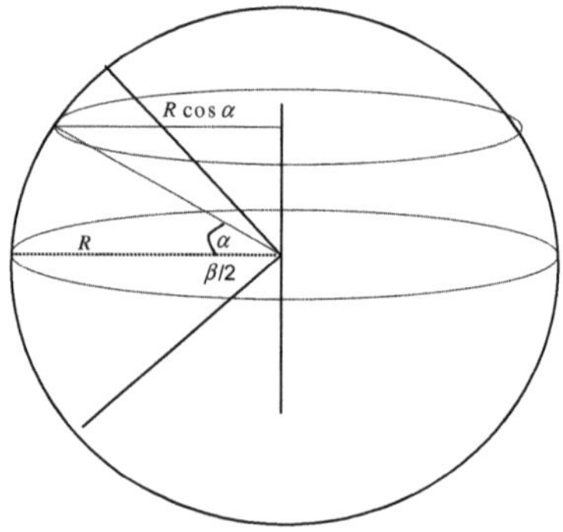

Figure 2: The dispersion area.

It follows that the density of fragments at range R is

$$\sigma_R = \frac{N}{4\pi R^2 \sin\frac{\beta}{2}}. \tag{10.2}$$

A person is represented in the model as a cylinder of diameter b and height cb, $c > 1$. The cylinder's exposed area to the fragments A is approximately a rectangle of width b and height that is determined by the dispersion angle β (see Figure 3).That is,

$$A = bMin\{2R\tan(\beta/2), cb\}. \tag{10.3}$$

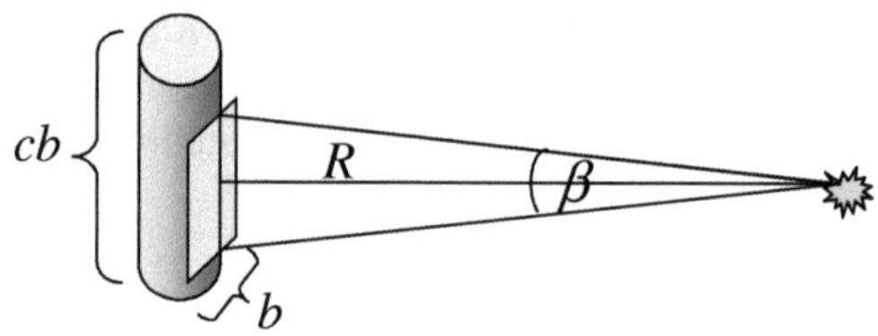

Figure 3: The exposed area.

Assuming uniform and independent dispersion of fragments, the probability that a person at range R is hit by at least one fragment is

$$P_H(R) = 1 - \left(1 - \frac{A}{4\pi R^2 \sin\frac{\beta}{2}}\right)^N. \tag{10.4}$$

From (10.2) it follows that

$$P_H(R) = 1 - \left(1 - \frac{A}{N/\sigma_R}\right)^N \cong 1 - e^{-A\sigma_R}. \tag{10.5}$$

We assume that the effectiveness of the explosive and the size of the arena are such that air resistance and gravitation have no significant effect on the trajectory and energy of the fragments. A fragment does not become harmless with distance.

10.2.2 Modeling the Arena

Some typical SB arenas, such as restaurants, may be depicted as a circular area around the SB. Let R_0 denote the distance between the SB (the EC belt) and the boundary of the arena. We assume that the crowd is distributed randomly and uniformly in the arena.

For modeling purposes, it is convenient to view the circular arena as a sequence of M concentric rings of width b. Each person occupies a round "slot" of diameter b in a certain ring. In particular, the SB is located in the central slot (see Figure 4a).

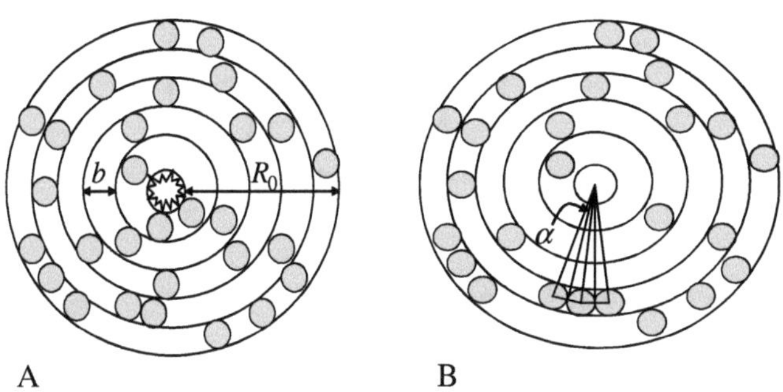

Figure 4: The arena.

The number of circles a_m that can fit into the mth ring is determined by the angle α_m between the line that connects the point of explosion with the center of a circle in ring m, and the tangent to that circle that passes through the point of explosion (see Figure 4b). That is, $a_m = \frac{2\pi}{2\alpha_m}$. But, $\sin\alpha_m = \frac{b/2}{mb} = \frac{1}{2m}$; therefore $\alpha_m = \arcsin(1/2m)$ and

$$a_m = \frac{\pi}{\arcsin(1/2m)}, \qquad m = 1,\ldots,M, \tag{10.6}$$

which is obviously independent of b. From now on we take $b = 1$.

It can be verified (at least for $M \le 200$, which is much more rings than we need for the SB scenario) that a_m is integer only for $m = 1$ ($a_1 = 6$). Assume that in each ring we pack, with possible small overlaps, $k_m = \lceil a_m \rceil$ slots – k_m is a_m rounded up. Define the *overlap factor* of ring m by

$$d_m = \frac{k_m - a_m}{k_m}. \tag{10.7}$$

For example, the overlap factors for m = 1, 2, 3, 4, 5, 6 are 0, 0.04, 0.01, 0.04, 0.02, 0.01, respectively. For higher values of M the d_m gets even smaller since the numerator in (6) is bounded by 1 and the denominator is (strictly) monotone increasing in m. See Figure 5 for m = 1, 2, and note the slight overlap of the circles in the second ring.

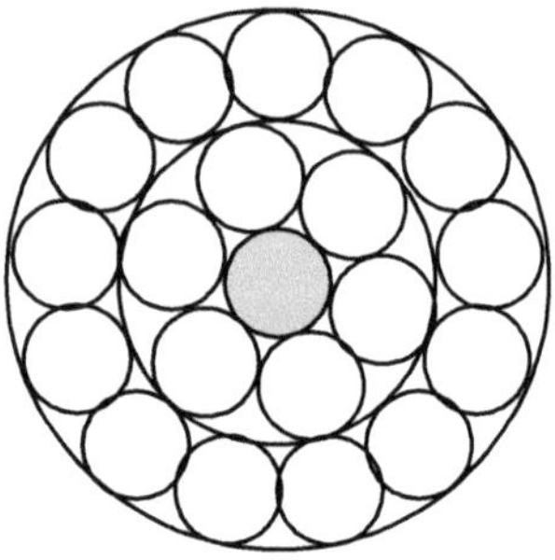

Figure 5: Number of slots, $k_1 = 6$, $k_2 = 13$.

Thus, the overlaps are marginal and they can be removed by slightly increasing the width of a ring. From now on we assume that the mth ring in the arena contains k_m slots.

Let,

$$K(m) = \sum_{n=1}^{m} k_n, \qquad m = 1,\ldots,M. \tag{10.8}$$

$K(M)$ is the maximum possible number of people in the arena (excluding the SB). For example, if $M = 10$, then $K(M) = 6+13+19+26+32+38+44+51+57+63 = 349$. Since we assume random homogeneous mixing in the arena, the expected number of people in the mth ring is

$$\mu_m = \frac{k_m}{K(M)} L, \tag{10.9}$$

where L is the number of people in the arena.

10.2.3 The Effect of Crowd Blocking

The number of casualties depends on the spatial distribution of people in the arena. Some people may become human shields to others by blocking the fragments. As shown in Figure 6, B1 totally shields (blocks) the target (T), and B2 partially shields it.

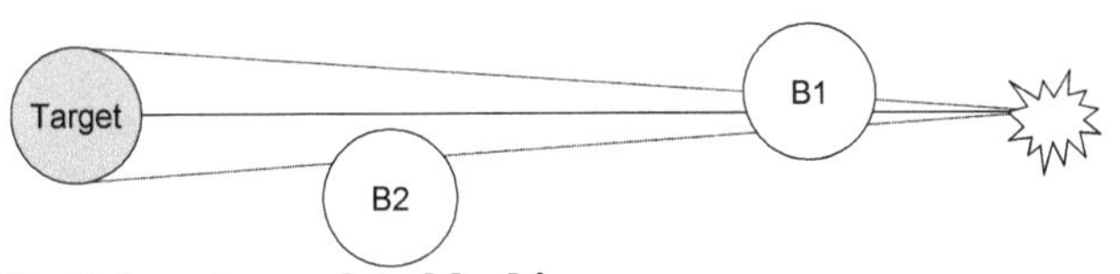

Figure 6: Partial and complete blocking.

A target (person) is assumed to be protected against a fragment that is moving its way if and only if the target's center is shielded from the beam spray. In other words, the target is safe if and only if there exists at least one person that blocks the line of sight (LOS) from the center of the SB's point of explosion to the center of the target. B1 in Figure 6 protects the target, while B2 does not. Instead of describing the situation in terms of persons, we will do it by looking at the slots in the various rings. Thus, a person (T) in ring m is protected if and only if there is at least one occupied slot in rings $1,\ldots, m-1$ that intersects the line of sight SB – T.

Since the slots in each ring are packed, there exists in each ring $1,\ldots, m-1$ exactly one slot that intersects the LOS SB – T. It follows that T is safe if and only if at least one of these $m-1$ slots is occupied by one of the other $L-1$ persons in the arena. Denote the probability of the complement of this event – the probability that a target in ring m is exposed and vulnerable to the explosion, given L people in the arena – by $\pi(L,m)$. Consider a certain person in ring m. Because of the uniform and independent distribution of people in the arena, the probability that the first of the other $L-1$ persons *will not* occupy a blocking slot is that he chooses one of $K(M) - m$ empty slots out of the $K(M) - 1$ available. The second person can choose from $K(M) - m - 1$ empty slots out of $K(M) - 2$ available, and so on. Thus,

$$\pi(L,m) = \begin{cases} \prod_{l=1}^{L-1}\left(1 - \frac{m-1}{K(M)-l}\right) & \text{if } L < K(M)-m+2 \\ 0 & \text{otherwise} \end{cases}, \tag{10.10}$$

and it is easily seen that $1 = \pi(L,1) \geq \pi(L,2) \geq,\ldots,\geq \pi(L,M)$.

Casualties that get directly hit by a fragment are called *primary casualties*. The expected number of primary casualties in the mth ring is

$$E_{L,m} = \mu_m \times \pi(L,m) \times P_H(m), \tag{10.11}$$

where from (10.5),

$$P_H(m) = 1 - e^{-A\sigma_m} = 1 - e^{-\frac{NMin\{2m\tan(\beta/2),c\}}{4\pi\sin\frac{\beta}{2}m^2}}. \tag{10.12}$$

The expected total number of primary casualties is

$$E_L(M) = \sum_{m=1}^{M} E_{L,m}. \tag{10.13}$$

Secondary casualties are persons located right behind exposed and hit persons. In terms of our model, person B (slot B) is susceptible to a secondary hit if there is exactly one slot A that blocks the LOS between the SB and B, and slot B is tangent to A. In other words, a person in ring m+1 might sustain a secondary hit if there is a person in ring m that stands between him and the SB. We assume that a secondary injury occurs independently, with probability q, provided the immediate shield has been hit. The situation where a secondary hit is possible in ring m happens when the slots in rings $m-1$ and m on the LOS are occupied, and slots $1,\ldots,m-2$ on the LOS are not. The event A = {*Slots* $1,\ldots,\ m-2$ *along the LOS are empty*} is the union of two disjoint events: B = {*Slots* $1,\ldots,\ m-1$ *along the LOS are empty*} and C = {*Slots* $1,\ldots,\ m-2$ *along the LOS are empty "and" slot* $m-1$ is occupied}. C is the event of interest and $P(C) = P(A) - P(B) = \pi(L,m-1) - \pi(L,m)$. Clearly, this event is possible only from the second ring on. Thus, a person in ring m is prone to primary hit with probability $\pi(L,m)$ and to secondary hit with probability $\pi(L,m-1) - \pi(L,m)$. The probability that this person is hit by a fragment is $\pi(L,m)P_H(m) + q(\pi(L,m-1) - \pi(L,m)P_H(m-1)$. It follows that the expected number $\hat{E}_{L,m}$ of primary and secondary casualties in ring m is

$$\hat{E}_{L,m} = \begin{cases} E_{L,m} & m = 1 \\ \mu_m[\pi(L,m)P_H(m) + q(\pi(L,m-1) - \pi(L,m)))P_H(m-1)] & \text{otherwise} \end{cases} . \tag{10.14}$$

10.2.4 Analysis

Figures 7 and 8 plot the values of $E_L(M)$ as functions of the crowd size L for the cases of perfect blocking (q = 0) and partial blocking ($q = 0.5$) . We examine two arena sizes, M = 10, 20, and two values of the spray beam angle, $\beta = 10^\circ$ and 60° . We assume that N = 100 fragments, and on average the height of a person is 3.5 times his width (c = 3.5); see Sheet "SB" in *Chapter10.xls*.

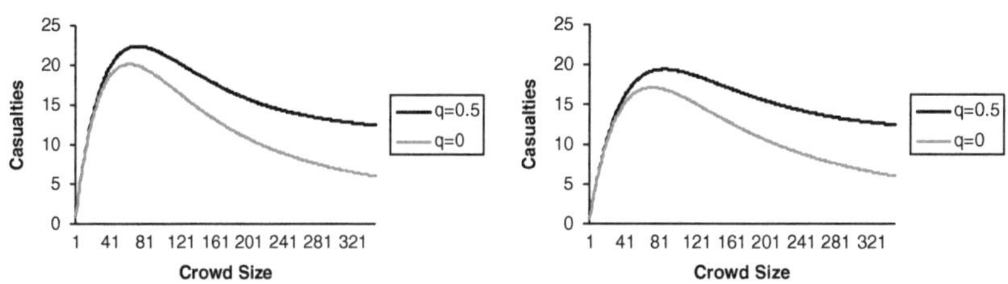

A B

Figure 7: Expected number of casualties. M = 10, $\beta = 10^\circ$ (A), 60° (B)

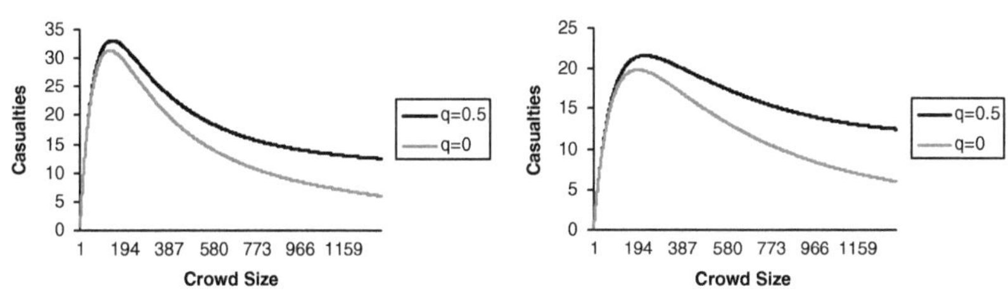

A B

Figure 8: Expected number of casualties. M = 20, $\beta = 10^\circ$ (A), 60° (B).

Figures 7 and 8 demonstrate that the effectiveness of the suicide bomb does not necessarily increase with the size of the crowd in the arena. Beyond a certain threshold, the expected number of casualties gets smaller. This phenomenon is attributed to crowd blocking, which becomes more significant as the density of the crowd increases. Note that in the case of perfect blocking ($q = 0$, primary casualties only) the expected number of casualties decreases to 6. This result is true in general since the effect of the explosion is limited to the first ring (with $k_1 = 6$) when the arena is fully crowded. The effectiveness of the explosion also depends on the size of the spray beam angle. When this beam is narrow, the effect is stronger than when it is wider. Note also that the addition of secondary casualties is significant only when the arena becomes crowded. In a low-density arena only direct hits of the fragments affect the number of casualties.

Figure 9 depicts the effect of the size of the arena on the expected number of casualties. We assume a crowd of $L = 100$ and we vary the size of the arena between $M = 6$ (the minimum size arena that can contain 100 people) and $M = 50$. We assume perfect blocking ($q = 0$).

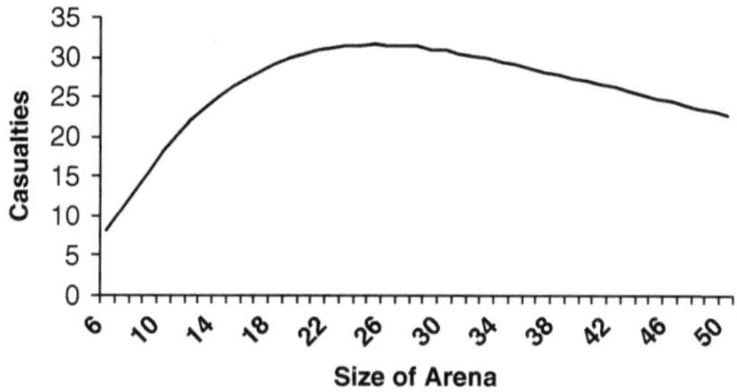

Figure 9: Expected number of casualties as a function of the arena size, $\beta = 10°$.

From Figure 9 we see that the size of the arena affects the damage in a non-monotone way. For a certain crowd ($L = 100$ in our example), there is a capacity ($M = 25$) for which the damage is maximal.

While the results shown in Figures 7–9 seem to be consistent with data regarding real SB events, a rigorous statistical analysis to confirm the numerical results of the model is difficult, if not impossible. First, there is no reliable record on the size of the crowd L in the arena at the time of the real event. This number can be estimated, at best, based on interviews of eyewitnesses. Second, while fragments are the main cause for fatalities in an SB event, some victims may have been killed by blast effects. Also, although some of the recorded injuries are related to mental shock and secondary injuries, such as cuts and bruises from debris, many of them may have been caused directly by the fragments. Third, the number of effective fragments in a suicide belt can also be only estimated. Fourth, the position of the SB in the arena affects the results, too. Incidents with relatively few casualties were typically consequences of a partial successful interdiction of the SB, where a guard identified the SB at the door, and as a result, the latter blew himself up outside the main part of the arena.

One possible takeaway from this model is that a natural thing to do in case of an imminent threat of an SB – run away – may not be the best strategy from the social welfare perspective. In some realistic situations such a response may actually increase the expected number of casualties. The instinct of the soldier who throws himself on a live hand grenade in order to protect his comrades is correct. Another takeaway is that such a model and analysis can help us recognize situations most attractive to an SB, and thus better take protective measures.

10.3 Response Policies for Bioterrorism – The Case of Smallpox

Responding to bioterror events that involve contagious agents is of major concern for authorities in their war against terror. There are many operational and logistic decisions that must be made in order to effectively cope with such a threat. These decisions are roughly divided into two levels: structural (strategic) decisions that are made in advance, before a possible bio-attack, and operational (real-time) decisions that must be made during such an event.

Some of the structural problems are as follows:

- decide how many vaccines to produce and stock,
- select the supply management policies for allocating, deploying, and controlling inventories of vaccine doses and other related supplies,
- set up the infrastructure (vaccination stations, quarantining facilities, transportation capacity, etc.),
- choose the vaccination procedure (e.g., inoculation only, pre-vaccination screening for contraindication),
- determine the manpower requirements and personnel assignment.

The operational (real-time) issues include

- detecting and identifying the type of the bioterror event,
- managing the contact tracing process (if applied),
- prioritizing efforts with respect to monitoring, isolating, quarantining, tracing, and vaccinating,
- coordinating the supply chain of vaccines and other supplies,
- identifying bottlenecks and potential congestion in the response process,
- determining service capacities and setting service rates.

One of the most critical decisions – a decision that has both structural and operational implications – is which vaccination policy to adopt. The vaccination policy decision has two levels. At the first level, policymakers must choose between essentially two options: a preemptive approach in which the entire population is pre-vaccinated, and a "wait-and-see" approach where post-attack emergency response (vaccination, curfew, isolation) commences following an outbreak of the disease. Mixtures of these two options are possible, i.e., pre-vaccination of first responders (e.g., health-care and law-enforcement personnel) only. Sociological and psychological considerations (is there a real threat or just a perceived one?) coupled with medical considerations (fear of vaccination side-effects) often hinder policy makers from taking any significant preemptive action.

If no significant preemptive measures are taken, the question at the second level is which post-event vaccination policy to adopt. The two main policies are *mass* vaccination and *trace* vaccination. In mass vaccination, maximum vaccination capacity is utilized to uniformly inoculate the entire population. In *trace* (also called *ring* or *targeted*) vaccination, only limited vaccination capacity is utilized to selectively inoculate contacts (or suspected contacts) of infected symptomatic individuals. In this section we develop a descriptive model for analyzing the effectiveness of mass vaccination for a highly contagious smallpox epidemic.

10.3.1 The Epidemic and Possible Interventions

Suppose that a terrorist releases smallpox viruses in a public area. The authorities are not aware of the event until a certain number of symptomatic patients are reported and diagnosed. Once the epidemic is detected and identified, a response is initiated, which may involve isolation, quarantine, tracing contacts of infected (and diagnosed) persons, and vaccinating some or all of the population. Smallpox has an incubation period during which an infected individual is asymptomatic and not contagious. The incubation period is divided into two periods of time: the *immunable* (also called *vaccine-sensitive*) period and the *non-immunable* period. We assume that vaccination is perfectly effective during the immunable period, and also for uninfected persons. During the non-immunable period, vaccination is not effective, and therefore the infected person will eventually become ill. Once the incubation period is over, the infected person becomes symptomatic and contagious, as long as the symptoms persist. Recovered patients of the disease are immune and not contagious.

Figure 10 presents the stages and progression of the disease. At any given time t, the population of non-vaccinated individuals is divided into six possible stages: S – susceptible to the disease; A – infected, not contagious, and immunable; B – infected, not contagious, and **not** immunable; I – infected, contagious, and not isolated; Q – infected, contagious, and isolated; R – removed (recovered, vaccinated, or dead). Figure 10 presents the transitions among the stages. Once the epidemic is identified, detected (symptomatic) infected individuals (in stage I) are isolated (moved to stage Q) and the vaccination process commences. Only asymptomatic persons, in stages S, A, and B, are vaccinated. While the vaccination is effective for individuals in stages S and A, it is ineffective for stage B, as shown in Figure 10. A detected symptomatic person (in stage I) is immediately isolated.

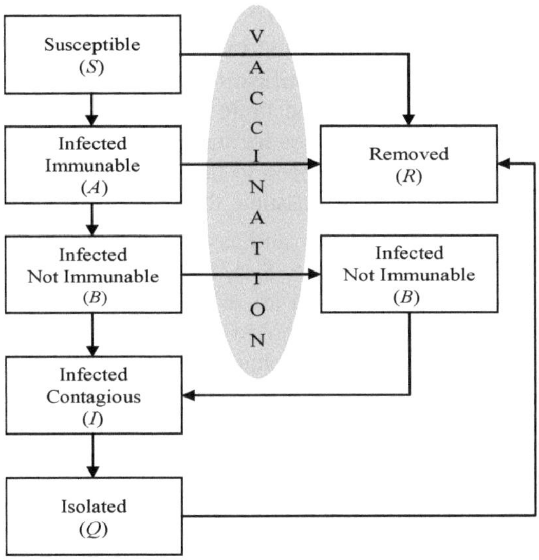

Figure 10: Stages and progression of the epidemic.

10.3.2 A Model for Mass Vaccination

The model for mass vaccination is based on the general SIR model for epidemics (Anderson and May, 1991) where only the stages *S*, *I*, and *R* are considered. From now on we use the same labels to indicate the name of the stage and the set of persons in that stage. The SIR model is a set of ordinary differential equations (ODE) similar to the deterministic Lanchester models of Chapter 5. However, in this chapter we present the models in their difference equations form. This form is more attune with the typical time resolution of a smallpox mass vaccination process – days – and is more manageable computationally

$$\begin{aligned} S(t+1) &= S(t)(1-\gamma I(t)), \\ I(t+1) &= I(t)(1+\gamma S(t)) - \rho), \\ R(t+1) &= R(t) + \rho I(t), \end{aligned} \tag{10.15}$$

where the parameter *t* represents days.

The number of newly infected depends on the interactions between susceptible *S* (from now on we drop the argument *t*, except for emphasis) and infected

I persons, which is measured by the product of these two variables. The parameter γ represents the intensity of the interactions. The probability that an infected person is removed in 1 day is ρ, so the average number removed is ρI. Thus, the rate of decrease in the number of susceptible persons is γSI, and the net change in the number of infected persons is $\gamma SI - \rho I$. We assume that the total population is fixed: $S + I + R = P$. We are employing expected value analysis (EVA) here, as described in Chapters 1 and 5, our justification being that the populations involved are mostly large. Results may be inaccurate when populations are small; in fact, (10.15) can even report negative values for state variables in some cases.

In the case of smallpox we also need to take into consideration the incubation and isolation stages. Thus, here we assume that $S + A + B + I + Q + R = P$. Let ω denote the daily vaccination capacity and assume perfect efficacy of the vaccination. Recall that only persons in stages *S, A*, and *B* are vaccinated. People in stage *I* who show up in a vaccination center are assumed to be processed (e.g., registration, screening) like the others, diagnosed and then isolated. Despite not being vaccinated, these persons consume vaccination capacity. Since the vaccination capacity is fixed at ω but the values of *S, A, B*, and *I* change over time, and since we assume that the vaccination queue is homogeneous in the sense that it reflects the distribution of stages in the population not yet isolated or removed, the probability that on a given day *t* a certain person in the population passes through a vaccination center is $V(t) = Min(1, \omega/(S(t) + A(t) + B(t) + I(t)))$. Thus, the number of persons in stage *I* that are isolated in the vaccination center is IV. A fraction λ of stage *I* persons that are *not* scheduled to appear in a vaccination center on a certain day show up voluntarily at a medical facility other than a vaccination center (family physician office, emergency room) are diagnosed and isolated. This is the *off-line* isolation process, where $\lambda I(1-V)$ infected individuals are being diagnosed and isolated before arriving at a vaccination center. Isolated persons are removed (die or recover) at a rate ρ. Let α and β denote the daily transition rates from stage *A* to *B*, and from *B* to *I*, respectively. The set of difference equations that describes the evolution of the smallpox mass vaccination process is

$$S(t+1) = S(t) \underbrace{- \gamma S(t) I(t)}_{\substack{\text{Susceptibles} \\ \text{that are infected}}} - \underbrace{S(t)V(t)}_{\substack{\text{Susceptibles that are} \\ \text{vaccinated}}}, \tag{10.16}$$

$$A(t+1) = \underbrace{\gamma S(t) I(t)}_{\substack{\text{Newly} \\ \text{infected}}} - \underbrace{\alpha A(t)}_{\substack{\text{Persons moving} \\ \text{from Stage } A \text{ to } B}} - \underbrace{A(t)V(t)}_{\substack{\text{Stage } A \text{ persons} \\ \text{that are vaccinated}}}, \tag{10.17}$$

$$B(t+1) = \underbrace{\alpha A(t)}_{\substack{\text{New persons} \\ \text{in Stage } B}} - \underbrace{\beta B(t)}_{\substack{\text{Persons that} \\ \text{move to Stage } I}}, \tag{10.18}$$

$$I(t+1) = \underbrace{\beta B(t)}_{\substack{\text{New persons} \\ \text{in Stage } I}} - \underbrace{I(t)V(t)}_{\substack{\text{Contagious persons isolated} \\ \text{via the vaccination process}}} - \underbrace{\lambda I\left(1 - V(t)\right)}_{\substack{\text{Contagious persons isolated before} \\ \text{reaching the vaccination center}}}, \tag{10.19}$$

$$Q(t+1) = \underbrace{I(t)V(t)}_{\substack{\text{Newly isolated persons}\\ \text{from vaccination centers}}} + \underbrace{\lambda I(t)\big(1-V(t)\big)}_{\substack{\text{Newly isolated persons not}\\ \text{from vaccination centers}}} - \underbrace{\rho Q(t)}_{\substack{\text{Persons removed}\\ \text{from isolation}}}, \tag{10.20}$$

$$R(t+1) = \underbrace{\rho Q(t)}_{\substack{\text{Persons removed}\\ \text{from isolation}}} + \underbrace{(S(t)+A(t))V(t)}_{\text{Effectively vaccinated persons}}. \tag{10.21}$$

The subpopulation S (10.16) decreases over time due to contraction of the disease and vaccination. The number of persons in the first stage of incubation (A, see (10.17)) increases due to infection and decreases by transiting to stage B or to stage R, due to vaccination. Equation (10.18) represents the dynamics at stage B; the set of people at this stage is fed by people from stage A and releases infected people, who become symptomatic, to stage I. The changes in the set of infected, symptomatic and contagious people I are shown in (10.19). New symptomatic and contagious people are generated from stage B, and existing such people are removed, either at the vaccination center or by reporting to medical facilities. The set of isolated people (10.20) increases at the rate of isolation (see (10.19)) and decreases by people who either recover or die. Finally, the "absorbing" stage R is fed by a flow of people removed from isolation and effectively vaccinated persons.

For more detailed modeling of smallpox vaccination processes, see Kaplan et al. (2002), Kress (2005b), and Kress (2006a).

10.3.3 Numerical Example

Consider a population of 10 million people. The number of initially infected by the bioterror attack is $A(0) = 1000$ people. The epidemic parameters (see Kress, 2005b, 2006a) are shown in Table 1. The difference Equations (10.16)–(10.21) are solved in sheet "MassVacc" in *Chapter10.xls*. Figures 11 and 12 show the evolution of the epidemic over time for various vaccination rates. Figure 11 presents the number of infected and contagious people (stage I) and Figure 12 presents the number of infected people in isolation (stage Q).

First we make the (optimistic) assumption that the authorities become aware of the bio-attack immediately, that is, the mass vaccination process is initiated immediately after the attack. In reality the bioterror attack would most likely be covert and therefore it will take a while until the first symptomatic patients (stage I) are detected (this situation is considered later).

Value	Definition	Symbol
10^{-7}	Infection rate	γ
0.33	Transition rate from stage *A* to *B*	α
0.087	Transition rate from stage *B* to *I*	β
0.33	Transition rate from stage *I* to *Q* (off vaccination process)	λ
0.083	Transition rate from stage *Q* to *R*	ρ

Table 1: The epidemic parameters (all rates are per day).

The newly infected stage *I* casualties are βB (see (10.19)). Figure 11 presents the newly infected and symptomatic people (Stage *I*), called henceforth – *casualties*, for three daily vaccination capacities: ω = 500K, 300K, and 100K.

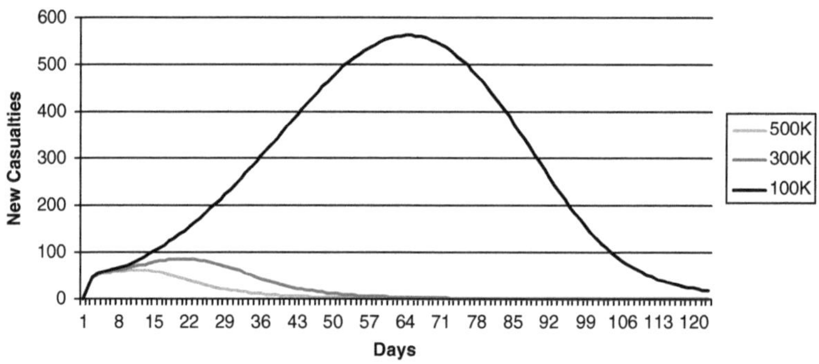

Figure 11: Newly infected casualties for three vaccination capacities.

The total numbers of casualties are 1508, 2823, and 33,576 for daily vaccination capacities 500K, 300K, and 100K, respectively. Clearly the effect is nonlinear. For example, if the daily vaccination capacity is reduced from 500K to 100K, the number of casualties increases by a factor of over 22. The peak of the epidemic occurs on day 12 (60 new casualties) when the daily vaccination capacity is 500K, and on days 21 (85 new casualties) and 65 (563 casualties) for daily vaccination capacities 300K and 100K, respectively. Also, the epidemic is eradicated after 63, 78, and 156 days for the three vaccination capacities, respectively.

An important insight from this model and analysis is logistical and relates to the question how big should be the isolation capacity. Figure 12, which looks very much like Figure 11, presents the number of casualties in isolation. The required isolation capacity is determined by the peak of these graphs. If the daily vaccination capacity is 500K then the isolation capacity needed is 589 beds. If it is 300K people per day, then the isolation capacity needed is 913 beds. However for daily vaccination capacity of 100K, the isolation capacity needed is 6266 beds.

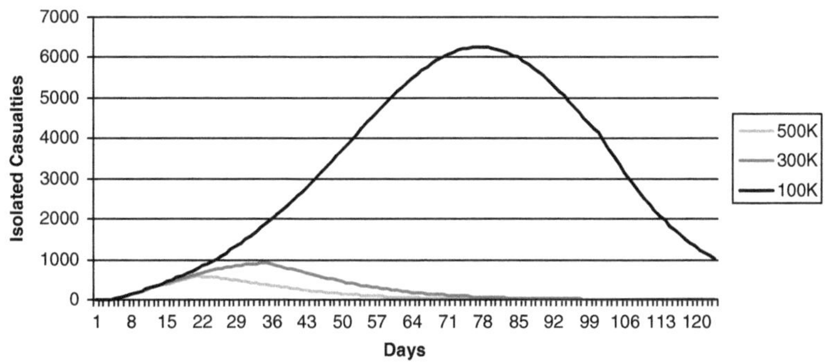

Figure 12: Casualties in isolation (Stage Q) for three vaccination capacities.

In reality the vaccination process will commence a few days after the attack, when the first symptomatic people show up in medical facilities and the authorities set up the vaccination process. Let d denote the time that elapses from the day of the attack to the day when the vaccination process commences. Figure 13 presents the number of newly infected for three delay scenarios: no delay, 4 days, and 8 days. Assume that the daily vaccination capacity is 500K. The total numbers of casualties are 1508 (see Figure 11), 2281, and 3289 for d = 0, 4, and 8 days, respectively.

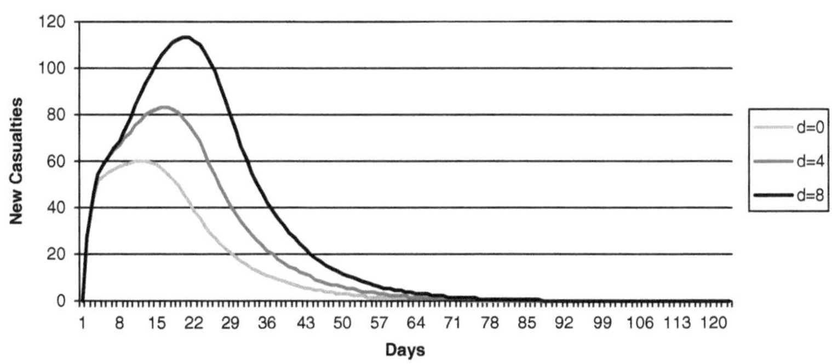

Figure 13: Newly infected casualties for three detection delay scenarios.

Notice that the delays in the epidemic detection have little effect on the duration of the epidemic. Figure 14 presents the time-dependent number of casualties in isolation.

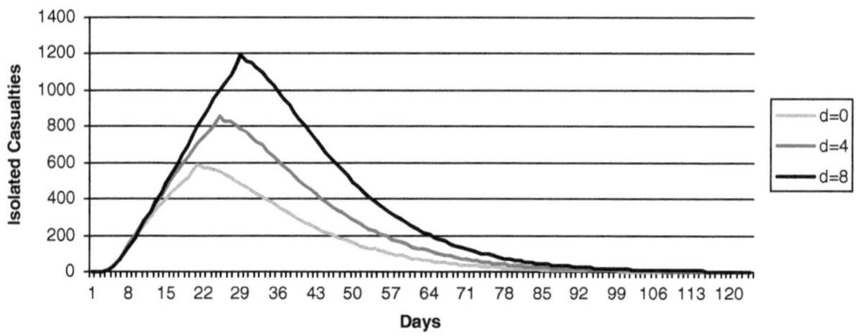

Figure 14: Casualties in isolation (Stage Q) for three detection delays.

The isolation capacity needed when $d = 8$ is more than twice the capacity needed when there is no delay.

10.4 Counterinsurgency

Insurgents take advantage of their small signature and elusiveness to launch attacks against the regime. Their effectiveness and ability to survive and operate depend on their ties with the general population. The insurgents get new recruits from the population and they rely on popular support for logistics and safety. As mentioned in the introduction, the key advantage of the insurgents is their elusiveness and invisibility. The regime forces, while having a significant advantage in terms of firepower and technology, simply cannot find the insurgents targets. As a result, not only may these targets evade the regime's attacks and continue their insurgency actions but also collateral damage caused to the general population from poor targeting by the regime's forces may generate an adverse response against the regime, and popular support for the insurgents. This popular support translates into new cadres of recruits to the insurgency. On the other hand, because of the large signature and exposure of the regime forces, the insurgents have perfect situational awareness and therefore their attacks against the regime forces are focused and effective. The insurgents also execute coercive actions against the population (e.g., suicide attacks, car bombs) with the objective to force the population to align with the insurgency. However, these actions may have mixed effects. The attitude of the population, which has a significant effect on the fate of the insurgency, is greatly affected by the sense of security (Lynn 2005, Hammes 2006). The population will align with the side that is perceived as better protecting it, or at least less threatening. Therefore, the coercive actions against the population by the insurgents may be a double-edged sword. On the one hand it may lead the population to believe that aligning with the insurgency will buy them protection against the insurgents' violent actions, but on the other hand it may instigate resistance to the insurgency and generate support for the regime who fights the insurgents. In this model we assume no such coercive actions by the insurgents. The main goal of the regime is eliminating the insurgency.

In the next section we present a dynamic model, which is a variant of a deterministic Lanchester system (see Section 5.2), for investigating cause-and-effect relations among the various parameters that affect the outcome of insurgency–counterinsurgency operations: force sizes, attrition rates, recruitment rates, popular support, and intelligence.

10.4.1 The Model

The model follows the dynamics of three groups: the insurgents I, the regime supporters in the population S, and the regime contrarians (insurgency supporters) in the population C. The fourth group – regime forces G – is assumed to remain fixed throughout the insurgency because the attrition the insurgents cause the regime forces does not affect significantly its size and capabilities. Thus, absent reinforcement, the regime force G, remains constant throughout the insurgency. A similar assumption is applied to the population; the attrition to the insurgency and the general population (as a result of collateral casualties) do not affect significantly the size of the population, which is typically several orders of magnitude larger than the attrition to P. Therefore we also assume that the total population $P = S + C + I$ remains constant.

The insurgents are diffused in the population and therefore, absent intelligence, the measure of their signature as targets for the regime force, is I/P, which may be interpreted as the probability that a randomly selected target is an insurgent. If the attrition coefficient of the regime forces is γ, then the attrition caused by the regime is γG (see Lanchester square law in Chapter 5), out of which a fraction I/P affects the insurgency and a fraction $(S+C)/P$ affects the general population. Thus, the attrition of the insurgency depends on both G and I (see the guerrilla warfare model in Chapter 5, and Deitchman (1962)). The insurgency recruits from the population of contrarians at a rate ρC. The dynamics of the insurgency is represented by the following ordinary differential equation (in the following $\dot{X}$ denotes the derivative of X with respect to time):

$$\dot{I} = -\gamma G\, I/P + \rho C. \tag{10.22}$$

The above equation assumes that the regime has no intelligence; it shoots in the dark and therefore it hits an insurgent with probability I/P. In reality, the regime force would invest effort in gathering intelligence, which leads to enhanced situational awareness. Denote the intelligence level by μ, $0 \le \mu \le 1$, where this parameter may be interpreted as the probability that the regime has good intelligence, which facilitates effective targeting. Thus, with probability μ the regime effort is focused on the insurgency targets and leads to a Lanchester square law (see Chapter 5) while with probability $1-\mu$ this effort is diffused and leads to the Deitchman guerrilla model (Deitchman 1962) mentioned above. Consequently, in the presence of partial intelligence, (10.22) becomes

$$\dot{I} = -\gamma G(\mu + (1-\mu) I/P) + \rho C\,, \tag{10.23}$$

and the rate of collateral casualties in the population is $\gamma G(1-\mu)(1-I/P)$.

There are several factors that affect the attitude of the population toward the regime. These include the economic situation, civil liberties, and political leadership. But, as mentioned above, the most significant factor is the sense of security. This model focuses on the security factor, and a simple way to represent this factor is by the weighted difference between the attrition the regime causes the insurgency, and the collateral casualties in the population, generated by the regime actions against the insurgency. This *weighted attrition balance* denoted by β is

$$\begin{aligned}\beta = \gamma G(\mu + (1-\mu) I/P) - v\gamma G(1-\mu)(1-I/P) = \\ \gamma G(\mu + (1-\mu)((1+v) I/P - v)).\end{aligned} \tag{10.24}$$

The weight parameter v represents the population's cognitive trade-off between insurgency casualties and collateral innocent civilian casualties. The larger the value of v, the more sensitive the population to its own casualties. If $\beta > 0$ then the attitude flow balance is directed from contrarians to supporters, and if $\beta < 0$, then the opposite is true. Notice that $\beta > 0$ if and only if $\mu/(1-\mu) > v - (1+v)I/P$. Assuming linear dependence on the attrition balance, and noting that $S = P - C - I$, we obtain that the supporters–contrarians dynamics in the population is captured by

$$\dot{C} = -Min\{\beta, 0\}\eta_S(P - C - I) - (Max\{\beta, 0\}\eta_C + \rho)C, \tag{10.25}$$

where η_C and η_S are the transition coefficients. The first term in (10.25) is the flow from S to C when $\beta < 0$. The second term represents outflow from C to S and I. The set of equations (10.23) and (10.25) describes the evolution of the insurgency. The variable S has been eliminated through the assumption that the total population P remains fixed.

10.4.2 Numerical Example

Consider an insurgency that starts off with $I_0 = 2000$ active insurgents. The regime force is G = 100,000. The population is P = 10^7 out of which initially 20% are contrarians and 80% are supporters of the regime. The values of the other parameters are summarized in Table 2. The time resolution of the model is days and the differential equations (10.23), (10.25) are solved as difference equations (see sheet "Counterinsurgency" in *Chapter10.xls*).

Value	Definition	Symbol
1.5×10^{-4}	Regime attrition	γ
0.5×10^{-4}	Insurgency recruitment	ρ
2	Population attrition trade-off	v
0.7	Intelligence level	μ
0.001	$S \to C$ transition	η_S
0.001	$C \to S$ transition	η_C

Table 2: The insurgency parameters.

The units of γ, ρ, η_S, and η_C are inverse time, while v and μ are dimensionless.

Note that at the beginning of the insurgency $2.33 = \mu/(1-\mu) > v-(1+v)I/P =$ $2-3\times2000/10^7 = 1.9994$, that is $\beta > 0$. Therefore, right at the beginning of the insurgency, more contrarians become supporters than the opposite and the opposition to the regime in the population fades away.

Figures 15 and 16 depict the evolution of the insurgency, over a period of time of 7 years, for three intelligence levels: 0.7 (base case – see Table 2), 0.5 (worse intelligence), and 0.9 (better intelligence). Figure 15 presents the insurgency and Figure 16 presents the number of contrarians in the population.

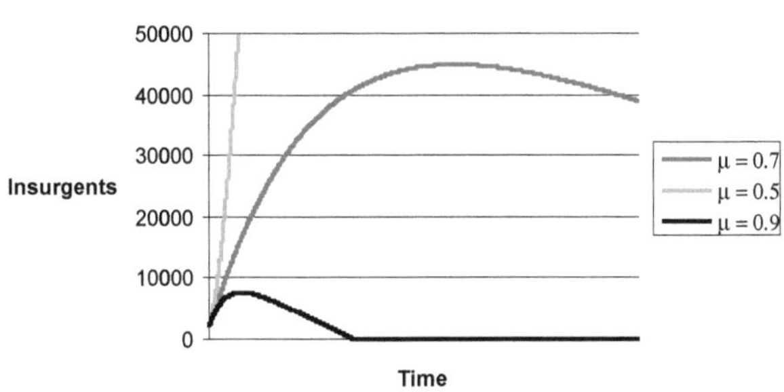

Figure 15: Size of the insurgency for three levels of intelligence.

We see that the fate of the insurgency is very sensitive to the level of intelligence. When the intelligence is poor $(\mu = 0.5)$ then $\beta < 0$ and the insurgency grows monotonically until the insurgents take over the regime. When $\mu = 0.7$, the insurgency grows for almost 4 years, reaches a size of over 40,000 insurgents, and then starts to decline. With very good intelligence $(\mu = 0.9)$ the insurgency reaches a maximum of just over 7500 insurgents after 190 days but is completely eradicated after less than 2.5 years.

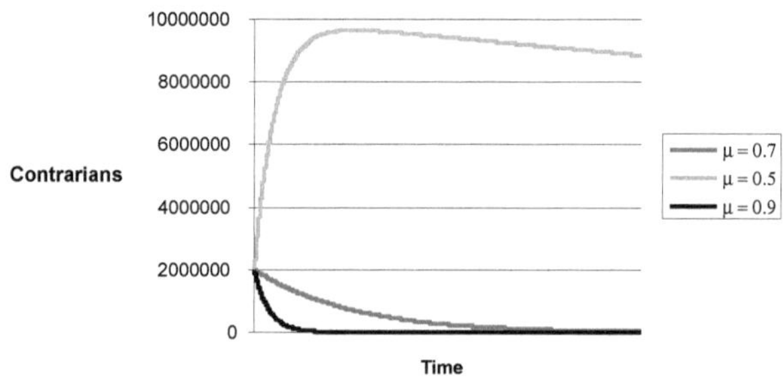

Figure 16: Number of contrarians for three levels of intelligence.

Similar behavior is observed with regard to the number of contrarians C in the population. When $\beta < 0$, C grows until there are no supporters left in the population. The slight decline in C after reaching its peak is due to ongoing recruitment to the insurgency. If $\mu \geq 0.7$, then $\beta > 0$, and the resistance to the regime in the population declines. However, when $\mu = 0.7$, the decline in the number of contrarians is not fast enough, therefore the insurgency can grow to over 40,000 (see Figure 15). Arguably, the recruitment coefficient ρ, which is fixed in our model, may be variable too; it may decrease when $\beta > 0$.

So far we assumed that the intelligence level μ is fixed. In reality it may change over time and improve as the regime succeeds in infiltrating the insurgency ranks and gathers more intelligence. Suppose that the intelligence level is a sigmoid function of time; it grows initially slowly, then grows rapidly, until it ebbs off when the regime reaches its maximum intelligence-gathering capabilities. Let, $\mu(t) = a/(1 + be^{-kt})$, where $\mu(0) = a/(1 + b)$ is the initial intelligence level, a is the maximum possible intelligence level, and k is a scale parameter that determines the evolution of $\mu(t)$ between the two extremes, see Figure 17. When μ is variable, the sign of β may change over time; while initially the insurgency grows for small values of μ, it starts to decline later on when μ gets larger. Figures 18 and 19 present the evolution of the insurgency and the number of contrarians, respectively, when μ behaves as in Figure 17.

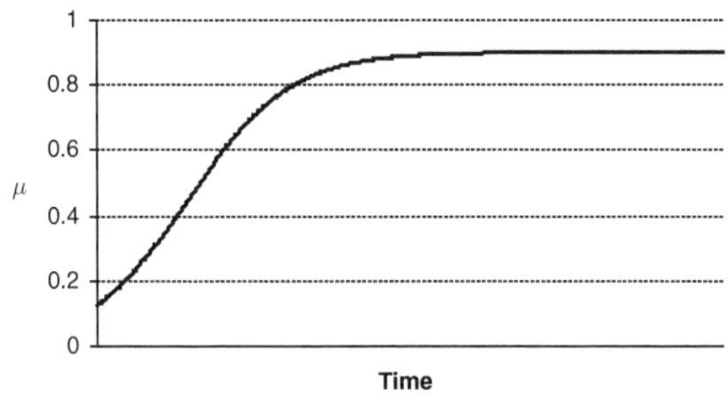

Figure 17: Intelligence level $\mu(t)$: a = 0.9, b = 6, k = 0.02.

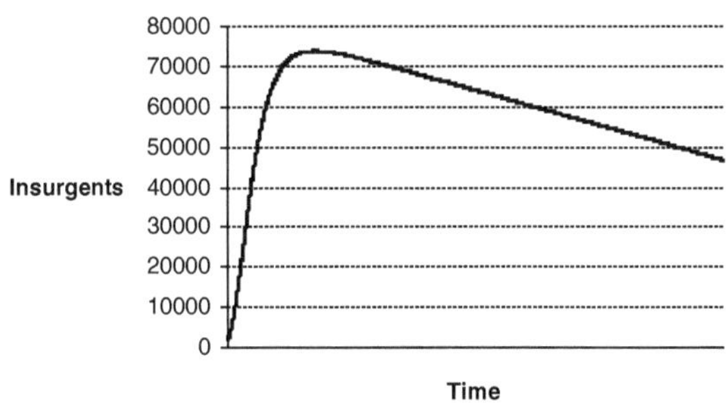

Figure 18: Size of insurgency for variable intelligence level.

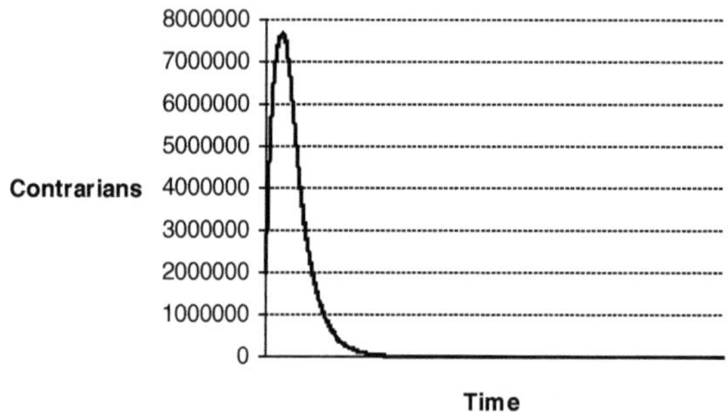

Figure 19: Number of contrarians for variable intelligence level.

The initial sharp increase in contrarians, followed by increase in insurgency, is due to initial poor intelligence ($\mu(0) = 0.13$). This trend is reversed when the intelligence is improved over time. The decline in C is much sharper than the decline in I because, according to our model, the insurgency decreases only because of attrition by the regime forces; it may increase due to popular support but it does not decrease solely because of lack thereof.

Exercises

(1) Consider the SB model in Section 10.2. There are $L = 50$ in an arena of size $M = 12$. Because of the nature of the event, the crowd is concentrated in one ring. Compute the expected number of casualties if
(a) the crowd is concentrated in ring $m = 9$
(b) the crowd is concentrated in ring $m = 12$ (the walls of the arena)
(c) the crowd is uniformly distributed in the arena, as in the model, and $q = 1$ (certain secondary injury)
Use sheet "SB" in *Chapter10.xls*.
Ans. (a) 25; (b) 16; (c) 17.8

(2) Consider once again the SB model. Suppose that instead of locating himself at the center of the arena the SB sets off the bomb at the boundary. Discuss qualitatively the effect of the attack in this case. Suppose that the arena comprises one ring and a center point $(M = 1)$. The crowd is four persons randomly distributed in the arena $(L = 4)$, absent blocking a person is hit with certainty $(P_H = 1)$, and there are no secondary casualties $(q = 0)$. For the data in Section 10.3.3, compare the expected number of casualties in the two scenarios: (a) the SB is at the center of the arena (see center part of Figure 5) and (b) the SB is in one of the six slots of the (outer) ring. Assume that a person is exposed to the SB only if the line connecting the centers of the two corresponding slots does not touch another occupied slot.

Ans. (a) 4, (b) $3\times\frac{7}{15}+2\times\frac{7}{15}+1\times\frac{1}{15}=2.4$

(3) Consider the bioterror model in Section 10.3. Obviously, timely detection of the attack and effective isolation process can reduce the number of casualties. Using sheet "MassVacc" in *Chapter10.xls*, plot a graph that presents the total number of casualties as a function of (a) the detection delay d and (b) the off-line isolation rate λ. Use the data in Section 10.3.3.

(4) Consider the numerical example in Section 10.4.2, and suppose that the intelligence level μ is constant as a function of time. Using sheet "Counterinsurgency" in *Chapter10*.xls answer the following questions: (a) what should be the minimum intelligence level such that the insurgency is eradicated after at most 2.5 years? (b) If $\mu = 0.8$, what is the minimum regime force size that will eradicate the insurgency within 2 years?
Ans. (a) 0.885; (b) 150,000.

Appendix A
Probability – the Mathematics of Uncertainty

The single most useful field of mathematics in the construction of combat models is mathematical probability theory. This appendix is an elementary introduction to the topic.

A.1 Preliminaries

Central to any discussion of probability is the idea of an observed experiment. We might flip a coin and observe whether it lands heads or tails, or shoot at a target and observe the miss distance, or pick up the *New York Times* and observe whether the word "Pirates" occurs in the headline. Events are things that either happen or do not when an experiment is performed. Symbols for events are customarily uppercase letters at the beginning of the alphabet, especially E. The probability of an event $P(E)$ is a measure of its likelihood when the experiment is performed. $P(E)$ is measured on a continuous scale [0,1] where 0 denotes impossibility and 1 denotes certainty.

It is important to be clear about what the experiment is. Let E be the event that certain developmental radar will detect a specified submarine periscope at a range of 5 miles in sea state 3. $P(E)$ can be just about anything, depending on details such as how often the periscope emerges, how long it stays up, and the altitude of the radar. Unless there is a clear understanding of the experiment, statements about probability are pretty much meaningless.

We often imagine that the experiment is repeatable, so that it can be performed over and over. The outcome of the experiment may vary with the replication, but the experiment still has the same chances of producing its various outcomes on every trial. In that case we can approximate the probability of a simple event, say the probability of a coin turning up heads, by counting the number of heads that occur in n tosses and dividing by the total number of tosses. As the number of tosses becomes larger and larger, this ratio approaches closer and closer to the precise value of the desired probability. In a similar manner, it would be possible to calculate the probabilities associated with the faces of a die, the probability of a newborn baby being a boy, and so on. Symbolically, the true probability of any such event is

$$P(E) = \lim_{n\to\infty} \frac{\text{number of times the event happens in } n \text{ trials of the experiment}}{n}. \quad (A.1)$$

Probability theory concerns the calculation of the probabilities of events in well-defined experiments. Probabilities are simply numbers, but events and experiments have yet to be given a mathematical form. It turns out that the most useful form is to imagine that an experiment always produces an outcome that be-

A. Washburn, M. Kress, *Combat Modeling*, International Series in Operations Research & Management Science 134, DOI 10.1007/978-1-4419-0790-5,

longs to a set called the sample space, and that any event is simply a subset of the sample space.

Example 1: Two navy battle groups, each comprising four ships, are under air attack.

Experiment: Counting the number of damaged ships after the attack.
Outcomes: (4,4), (4,3), (4,2), (4,1), (4,0), (3,4),…,(0,0).
Sample Space: Comprises 25 outcomes $\{(i,j)\ i\ j = 0,1,2,3,4\}$ ({..} – symbol of a set).
Event: E = {Total number of damaged ships is no more than 2} = {(0,0), (0,1), (1,0), (1,1), (2,0), (0,2)}.

Example 2: Engage a target with five rounds of fire.

Experiment: Observe the result of a shot. Stop after the first hit or when ammunition is used up.
Outcome: A sequence of misses (M) that may or may not end with a hit (H).
Sample Space: {H, MH, MMH, MMMH, MMMMH, MMMMM}.
Event: E = {The target has not been hit} = {MMMMM}.

We have seen that an event is associated with a set of outcomes. Sets can intersect or unite. The *intersection* of two sets A and B, denoted as $A \cap B$, is the set of all the outcomes that are *both* in A and B. Consider Example 1. Let A be the event "Battle Group 2 lost one ship = {(0,1), (1,1), (2,1), (3,1), (4,1)}, and let B be the event "Battle Group 1 lost two ships" = {(2,0), (2,1), (2,2), (2,3), (2,4)}. The intersection is the event "Battle Group 2 lost one ship *and* Battle Group 1 lost two ships" = {(2,1)}, the only outcome in both A *and* B. The *union* of two sets A and B, denoted as $A \cup B$, is the set of all the outcomes that are *either* in A *or* in B. The *union* of the events A and B defined above is the event "Battle Group 2 lost one ship *or* Battle Group 1 lost two ships" = {(0,1), (1,1), (2,1), (3,1), (4,1) {(2,0), (2,2), (2,3), (2,4)}. The complement of a set A, denoted as $\overline{A}$ (or A^C), is the set of all the outcomes in the sample space that are *not* in A. For example, the complement of the event C = "Battle Group 1 lost at least two ships" = {(2,0), (2,1), (2,2), (2,3), (2,4), (3,0), (3,1), (3,2), (3,3), (3,4), (4,0), (4,1), (4,2), (4,3), (4,4)} is $\overline{C}$ = "Battle Group 1 lost at most one ship" = {(0,0), (0,1), (0,2), (0,3), (0,4), (1,0), (1,1), (1,2), (1,3), (1,4)}. The union of C and $\overline{C}$ is the entire sample space, denoted by S.

A.2 Computing Probability

The probability of an event depends on the composition of outcomes in its corresponding set. If all outcomes in the sample space are equally likely to occur, then the probability is simply the number of outcomes in that set divided by the number of outcomes in the sample space.

Example 3: A fair die is tossed once. The sample space is $S = \{1, 2, 3, 4, 5, 6\}$ and because the die is fair all six outcomes are equally likely. The probability of the event E = "the outcome is an even number" is

$$P(E) = \frac{\text{Number of outcomes in } \{2, 4, 6\}}{\text{Number of outcomes in } S} = \frac{3}{6} = \frac{1}{2}. \tag{A.2}$$

Using the concepts and definitions discussed thus far, the following basic relationships can be derived for any two events A and B:

- The probability of an event is between 0 and 1: $0 \le P(A) \le 1$.
- $P(\bar{A}) = 1 - P(A)$. The sum of probabilities of an event and its complement is 1.
- $P(A \cap \bar{A}) = P(\phi) = 0$ (ϕ is the empty set). It is impossible that an event and its complement will occur together.
- $P(A \cup \bar{A}) = P(S) = 1$. Either an event or its complement must occur.
- $P(A \cup B) = P(A) + P(B) - P(A \cap B)$. Implication: if A and B are disjoint (no outcomes in common) then $P(A \cup B) = P(A) + P(B)$.

Example 4: A fair die is tossed once. A = "the outcome is an odd number" = $\{1, 3, 5\}$, B = "the outcome is smaller than 4" = $\{1, 2, 3\}$. P(the outcome is odd number that is smaller than 4) = $P(A \cap B)$ = Number of outcomes in the set $\{1, 3\}$/Number of outcomes in S = 2/6 = 1/3. $\bar{A}$ = "the outcome is not an odd number"= $\{2, 4, 6\}$. $P(A \cup \bar{A})$ = P("outcome is either an odd number or an even number") = $P(S)$ = 1.

A.3 Conditional Probability and Independence

The probability that two events, A and B, occur at the same time is the probability of the *intersection* of the two corresponding subsets of the sample space, as shown above. In this section we will compute this probability when the events are *independent*. First, however, the notion of *conditional probability* will be discussed.

The conditional probability of one event, B, occurring given that another event, A, has occurred is denoted by $P(B|A)$. The meaning of the words *condition* and *given* is that essentially the sample space S has "shrunk" to the set A. The term "Given A occurred" means that only outcomes that belong to A matter now; the other outcomes in S may be ignored because we know that they cannot occur simply because A has occurred. Thus, the probability $P(B|A)$ is the number of outcomes in B that are also in A, divided by the number of outcomes in A (which currently is our shrunk sample space). If $n(A \cap B), n(B)$ and $n(S)$ denote the number of outcomes that are both in A and B, in B, and in the entire (original) sample space, before it has been shrunk into A, then,

$$P(B|A) = \frac{n(A \cap B)}{n(A)} = \frac{n(A \cap B)/n(S)}{n(A)/n(S)} = \frac{P(A \cap B)}{P(A)}. \qquad \text{(A.3)}$$

Equation (A.3) is very important; it leads to many other properties and useful results. We can see right away that

$$P(A \cap B) = P(B|A)P(A). \qquad \text{(A.4)}$$

The expression in (A.4) is known as the *multiplicative law* of probability.

Example 5: A fair die is tossed once. What is the conditional probability that the outcome is 2 given that the outcome is even? $A = \{2,4,6\}$, $B = \{2\}$, $A \cap B = 2$. $P(B|A) = P(A \cap B)/P(A) = (1/6)/(3/6) = 1/3$.

The events A and B are said to be *independent* if the conditional probability of event B given event A equals the unconditional probability of event B. In other words, whether or not A is known to occur has no bearing on the probability of B's occurrence. In this case, $P(B|A) = P(B)$ and the multiplication law simplifies to $P(A \cap B) = P(A)P(B)$.

Example 6: When we say that two rounds of fire are independent of each other we mean that the outcome of one shot does not alter the probability of success of the other shot. In other words P(Shot 2 hits | Shot 1 hits) = P(shot 2 hits) and therefore P(Shot 1 hits and Shot 2 hits) = P(Shot 1 hits) x P(Shot 2 hits).

A.4 Bayes Rule

Suppose that in some experiment the event B can occur only in conjunction with one of several mutually exclusive events $A_1, \ldots, A_n$. We know the probabilities $P(A_i),\ i = 1, \ldots, n$ and the conditional probability $P(B|A_i),\ i = 1, \ldots, n,$ and that B occurs given that A_i has occurred. Then the *theorem of total probability* allows us to calculate the unconditional probability of the event B:

$$P(B) = P(B|A_1)P(A_1) + \ldots + P(B|A_n)P(A_n). \qquad \text{(A.5)}$$

Bayes Rule (or Bayes theorem) uses (A.5) and the definition of conditional probability to reverse the condition between B and A_i. The conditional probability of A_i given B is

$$P(A_i|B) = \frac{P(B|A_i)P(A_i)}{P(B|A_1)P(A_1) + \cdots + P(B|A_n)P(A_n)}. \qquad \text{(A.6)}$$

Example 7: There are three boxes, 1, 2, and 3. Each box contains 10 items of a certain product. There are 5, 6, and 8 good items in boxes 1, 2, and 3, respectively. The remaining items are defective. A box is chosen at random and then an item is drawn from it randomly. If the item drawn is defective, what is the probability that box A was chosen? To solve let B = {item is defective} and A_j = {box j is selected}, j = 1, 2, 3. Clearly $P(A_j)$ = 1/3, j = 1, 2, 3. Also, $P(B|A_1) = 5/10$, $P(B|A_2) = 4/10$, and $P(B|A_3) = 2/10$. The desired probability, using (A.6) with $n = 3$, is

$$P(A_1 \mid B) = \frac{(5/10)(1/3)}{(5/10)(1/3) + (4/10)(1/3) + (2/10)(1/3)} = \frac{5}{11}.$$

Example 8: The probability that a certain object in a target area is a valuable target is 0.3. An imperfect sensor observes the object and declares it to be a valuable target. The probability the sensor fails to recognize a valuable target is 0.2, and the probability it erroneously identifies a worthless target as valuable is 0.3. What is the probability that the object is a valuable target? Let V = {the object is a valuable target} and V^C = {the object is not a valuable target}. Let SV denote the event that the sensor identifies the object as valuable target and SV^C the event that the sensor identifies the object as not valuable. We wish to compute $P(V|SV)$:

$$P(V \mid SV) = \frac{P(SV \mid V)\,P(V)}{P(SV \mid V)\,P(V) + P(SV \mid V^C)\,P(V^C)}$$

$$= \frac{(0.8)(0.3)}{(0.8)(0.3) + (0.3)(0.7)} = 0.53.$$

A.5 Random Variables

A random variable is a transformation which associates with each point or collection of points in the sample space some real number. For example, in the case of two dice being tossed simultaneously, the outcomes could be represented very conveniently by a random variable, X, where X = the *sum* of the spots on the dice (possible values: 2,…,12). Other possible random variables associated with the sample space generated by tossing two dice are the *absolute difference* between the spots on the two dice (possible values: 0,…,5), the *greater of the two numbers* of spots showing (possible values: 1,…,6), the *number of 3's* showing (possible values: 0, 1, 2), or whether the outcome is either 7 or 11 (possible values: 1 if either of those outcomes results, otherwise 0). In each case, the random variable specifies a *number* to be associated with each possible outcome of the experiment. A random variable is said to be *discrete* if it can take on only a finite or countably infinite number of values. Otherwise it is said to be *continuous* and usually takes on as values all the real numbers in some interval.

A random variable is represented by two related functions that describe its probabilistic behavior. For a discrete random variable these two functions are called *probability mass function* (pmf) (or in short, *probability function*) and *cumulative distribution* function (cdf). The pmf $p_X(x)$ is simply the probability that the value of a random variable X is x: $p_X(x) = P(X = x)$. The cdf is defined as $F_X(x) = P(X \le x)$. If there is no danger of confusion about which random variable is being described, the subscript X will be omitted in both cases.

Example 9: A dice is tossed twice. Let X = sum of the spots. Then $p(5) = 4/36 = 1/9$ and $p(12) = 1/36$. Also $F(4) = p(2) + p(3) + p(4) = 1/36 + 2/36 + 3/36 = 1/6$.

The probability associated with a continuous random variable is characterized by a function f called a *probability density function* (pdf) or simply *density function.* The area under this function represents probability and hence the total area under this curve is 1. The area $f(x)\Delta x$ gives the approximate probability that the value of the random variable X will occur in an interval of length Δx around x. The cdf for a continuous random variable is

$$F(x) = P(X \le x) = \int_{-\infty}^{x} f(x)dx. \tag{A.7}$$

It follows that $\lim_{x\to-\infty} F(x) = 0$ and $\lim_{x\to\infty} F(x) = 1$, and

$$P\{x_1 \le X \le x_2\} = \int_{x_1}^{x_2} f(x)dx = \int_{-\infty}^{x_2} f(x)dx - \int_{-\infty}^{x_1} f(x)dx = F(x_2) - F(x_1). \tag{A.8}$$

The pdf of a continuous random variable is obtained by differentiating the cdf.

Example 10: The distance between two adjacent mines is 1000 m. A ship can cross the line connecting the two mines anywhere on that line with equal probability and therefore any distance between 0 and 500 m from a mine is equally likely. Thus, $f(x) = 1/500$. The probability that the distance of the ship to the nearest mine is between 200 and 300 m is given by $P(200 \le X \le 300) = \int_0^{300} \frac{1}{500} dx - \int_0^{200} \frac{1}{500} dx = \frac{100}{500} = .2.$

The general term probability distribution is used to denote the particular probability functions applicable to a given random variable. In the discrete case, the probability distribution may be specified by stating either the pmf or the cdf. The discrete probability distributions encountered most frequently in this text are the *binomial* distribution, the *Poisson* distribution, and the *geometric* distribution, each of which is discussed separately later on in this appendix. A continuous probability distribution may be specified by identifying either its pdf or its cdf. Among the most common of these continuous probability distributions are the *uniform* (see Example 10), *normal,* and *exponential* distributions. These and the *bivariate normal* distribution, which is an example of a continuous joint distribution of two variables, are also discussed in this appendix.

Similar to the definition of independence with respect to events (Section A.3), two random variables X and Y are said to be *independent* if $P(X \le x \text{ and } Y \le y) = P(X \le x)P(Y \le y) = F_X(x)F_Y(y)$.

A.6 The Mean and Variance of a Random Variable

As the name implies, the *mean*, also called *expected value*, of a random variable is a measure of "average value" of the random variable. It is somewhat analogous to the center of gravity used in physics. The symbols commonly used to represent the mean of a random variable X are $E[X]$ and μ. In the discrete case, the mean is calculated by taking the weighted average of all the possible values of the random variable, i.e., by multiplying each value by its respective probability of occurrence and then summing all of the resulting products:

$$E(X) = \mu = \sum_{\text{All possible } x_i} x_i p(x_i)\,. \tag{A.9}$$

Example 11: The number of targets X that are present in a certain target area is x with probability $p(x) = 0.5^{x+1}/(1-0.5^4)$, $x = 0, 1, 2, 3$. The expected value of the number of targets is

$$E(X) = 0 \times \frac{8}{15} + 1 \times \frac{4}{15} + 2 \times \frac{2}{15} + 3 \times \frac{1}{15} = \frac{11}{15}.$$

The extension of the concept of mean value to the continuous case follows directly in a manner completely analogous to the discrete case. The only difference is that the sum is replaced by an integral

$$E(X) = \mu = \int_{\text{All possible } x} x f(x) dx. \tag{A.10}$$

Example 12: Consider Example 10. The expected distance between the ship and the nearest mine is

$$E(X) = \int_0^{500} \frac{x}{500} dx = 250 \text{ m, as one would expect.}$$

The expected value is a *linear operator*, that is, if a and b are any constants, then $E(aX+b) = aE(X)+b.$ Furthermore, for any set of random variables (dependent or independent) $X_1, \ldots, X_n$, we have that $E(\sum_{i=1}^{n} X_i) = \sum_{i=1}^{n} E(X_i)$.

The *variance* of a random variable is a measure of its dispersion from the mean. Formally,

$$V(X) = \sigma^2 = E(X-\mu)^2. \tag{A.11}$$

A useful formula for computing the variance is obtained by observing that

$$E(X-\mu)^2 = E(X^2 - 2\mu X + \mu^2) = E(X^2) - 2\mu E(X) + \mu^2 = E(X^2) - \mu^2. \quad \text{(A.12)}$$

To illustrate the usefulness of the variance as a measure of dispersion, it is convenient to introduce a unit or measure called the *standard deviation*, σ, which is defined as the square root of the variance. A probability distribution having a small σ^2 and hence a small σ would be concentrated tightly about the mean. On the other hand, a probability distribution having a large σ would be dispersed more widely about the mean.

Example 13: Consider Example 10. The variance of the distance between the ship and the nearest mine is $V(X) = E(X^2) - \mu^2 = \int_0^{500} \frac{x^2}{500} dx - 250^2 = \frac{500^3}{3 \cdot 500} - 250^2 = 20833.33$ m. The standard deviation of this distance is $\sigma = \sqrt{V(X)} = \sqrt{20833.33} = 144.34$ m.

It is easily shown that $V(aX+b) = a^2 V(X)$. Note that adding a constant to a random variable does not change its variance at all. If $X_1, ..., X_n$ are independent random variables, then $V(\sum_{i=1}^{n} X_i) = \sum_{i=1}^{n} V(X_i)$. This relation does not hold if the random variables are not independent. Note that it is variances that add when independent random variables are summed, not standard deviations. Adding standard deviations is almost always the wrong thing to do. For example, suppose a shot misses a target for two independent reasons, one that the gun is aimed at the wrong place (error *X*) and the other that the wind blows the shot around as it moves toward the target (error *Y*). The two errors have means of 0 and standard deviations 30 and 40 m, respectively. The total error is *X*+*Y*, but the standard deviation of that error is not 70 m, but rather 50 m, the square root of the sum of 900 and 1600 m^2. Variances add, not standard deviations.

The *correlation coefficient* of two random variables *X* and *Y* is

$$\rho = \rho(X,Y) = \frac{E[(X-\mu_X)(Y-\mu_Y)]}{\sigma_X \sigma_Y} = \frac{E(XY) - E(X)E(Y)}{\sigma_X \sigma_Y}, \quad \text{(A.13)}$$

where μ_X and μ_Y are the mean values of *X* and *Y*, respectively, and σ_X and σ_Y are the standard deviations, respectively. If the two random variables are independent then $\rho = 0$. If $\rho \neq 0$ then the two random variables are said to be *correlated.*

A.7 Discrete Probability Distributions

In this section we describe certain widely used discrete probability distributions: binomial, geometric, and Poisson.

Binomial Distribution

The binomial distribution arises from a sequence of simple independent trials in which the occurrence or the nonoccurrence of an event is the only item of interest. If, in such a sequence of trials, the individual probability of the event's occurring on any one trial remains constant through the sequence, then these trials are referred to as *Bernoulli* trials. A good example of Bernoulli trials is a sequence of coin tosses for which the probability of a *head* on a single toss is p and the probability of a *tail* is $q = 1 - p$. Another example is when each of a sequence of targets is engaged by exactly one round of fire. The shots are independent and the targets are equally vulnerable; the probability a target is killed following a shot is a constant p, where the probability it is still alive is $q = 1 - p$. In situations such as these, it is desirable to know the probability of a specified number of occurrences of the event in a given number of trials (e.g., the number of killed targets following n shots). An occurrence of the event is sometimes referred to as a *success* and a nonoccurrence as a *failure*. The binomial distribution provides the means of determining the probability of exactly x occurrences of an event in n independent trials. The probability mass function of the binomial distribution is

$$P(X = x) = \binom{n}{x} p^x q^{n-x}, \quad x = 0, 1, \ldots, n, \tag{A.14}$$

where n is the number of trials, x is the number of successes, and $\binom{n}{x} = \frac{n!}{x!(n-x)!}$, the number of combinations of n things taken x at a time. The expected number of successes is $\mu = np$ and the variance is $V(X) = \sigma^2 = npq$.

Example 14: A shooter, having five rounds of fire, engages five targets, each with one round of fire. The probability that a round of fire kills a target is 0.7. The probability that exactly three targets are killed is $P(X = 3) = \binom{5}{3}(0.7)^3(0.3)^2 = 0.309$. The probability that at least four targets are killed is $P(X \geq 4) = \binom{5}{4}(0.7)^4(0.3)^1 + \binom{5}{5}(0.7)^5(0.3)^0 = 0.528$. The expected number of killed targets is $\mu = 5 \times (0.7) = 3.5$.

Geometric Distribution

It has been seen that the binomial distribution applies to situations where the random variable is the number of successes in n independent Bernoulli trials with a constant probability p of success on each trial. An associated random variable is the number of independent Bernoulli trials until the first success occurs. The probability of this number is given by the *geometric distribution* and its pmf is

$$P(X = x) = (1-p)^{x-1} p, \quad x = 1, 2, \ldots . \tag{A.15}$$

The expected number of trials until the first success is $\frac{1}{p}$ and the variance is $V(X) = \sigma^2 = \frac{(1-p)}{p^2}$.

Example 15: A shooter repeatedly engages a target until it is killed. The probability that a round of fire kills a target is 0.7 and the shots are independent. The probability that the shooter will shoot no more than two rounds until his mission is over is $P(X \le 2) = p + (1-p)p = 0.7 + 0.3 \times 0.7 = 0.91$. The expected number of shots until the target is killed is 1/0.7 = 1.43 rounds.

Poisson Distribution

Another extremely useful probability distribution is the *Poisson*, which typically applies in situations in which an event occurs repeatedly in a "completely random" or "haphazard" manner. A random variable X, which is Poisson distributed is the *number* of such occurrences in a certain time interval. The probability function for a Poisson random variable with parameter λ is defined by

$$P(X = x) = \frac{\lambda^x e^{-\lambda}}{x!}, \quad x = 0, 1, 2, \ldots . \tag{A.16}$$

A distinctive property of the Poisson distribution is that its mean equals its variance, specifically, $\mu = \sigma^2 = \lambda$. The Poisson distribution approximates the binomial distribution when n is large, p is small, and $\lambda = np$. Another important property of the Poisson distribution is that a sum of independent Poisson random variables has also a Poisson distribution. Specifically, if each $X_i, i = 1, \ldots, n,$ has a Poisson distribution with parameter λ_i, and $X_1, \ldots, X_n$ are independent, then $\sum_{i=1}^{n} X_i$ has also a Poisson distribution with parameter $\bar{\lambda} = \sum_{i=1}^{n} \lambda_i$.

Example 16: The number of emissions from a radioactive material is a Poisson random variable with parameter $\lambda = 20t$, where t is the time in hours. The probability that exactly five emissions occur in 1/2 h is derived by first computing the parameter (mean and variance) λ that corresponds to 1/2 h:

(20)(1/2) = 10 emissions. The probability of five emissions in 1/2 h is $P(X=5)=\frac{10^5 e^{-10}}{5!}=0.038.$

A.8 Continuous Probability Distributions

In this section we describe certain widely used continuous probability distributions, where the random variable can obtain any real value in a certain interval. The distributions described in the following are uniform, exponential, and normal, including the bivariate normal.

Uniform Distribution

Consider an interval $[a,b]$ in which each point is equally likely to be selected. Thus, the probability that a selected point in that interval is in the sub-interval $[a,x]$, $a \le x \le b$, is proportional to the length of $[a,x]$. Thus, the cumulative distribution function is

$$F(x)=P(X\le x)=\frac{x-a}{b-a},\quad a\le x\le b. \tag{A.17}$$

The density function is obtained by differentiating $F(x)$: $f(x)=\frac{1}{b-a}$. The mean of the uniform distribution with parameters (boundaries) a and b is $\mu=\frac{a+b}{2}$. The variance is $\sigma^2=\frac{(b-a)^2}{12}$. See Examples 10 and 13.

Example 17: A target can be present anywhere on the interval [0,1]. The probability the target is located between the points 0.4 and 0.7 is $P(0.4\le X\le 0.7)=0.7-0.4=0.3$. The mean point is 0.5.

Exponential Distribution

The *exponential distribution* has only one parameter λ. This distribution is probably the most widely used in operations research. The cdf of the exponential distribution is

$$F(x)=1-e^{-\lambda x},\quad x\ge 0 \tag{A.18}$$

and its pdf is $f(x)=\lambda e^{-\lambda x}$. The mean of this distribution is $1/\lambda$ and the variance is $1/\lambda^2$. An important property of the exponential distribution is that it is *memoryless*; the probability that the next event will occur, say, in the next 10 s is independent of the time elapsed since the last event. An exponential random variable is often used to model the (next) time of occurrence of "random" or "haphazard" events, e.g., the time it takes a "no-wear" part to fail, for a radioactive isotope to emit a radiation blast or a customer to arrive at a bank. This distribution plays a key role in the Poisson process, described later on.

Example 18: The time between two consecutive messages arriving at the intelligence cell in a brigade command post follows an exponential distribution with mean of 5 min. The probability that during a period of 7 min there is no message is $P(X \geq 7) = 1 - F(7) = e^{-\frac{7}{5}} = 0.25$.

An interesting property of the exponential distribution is that the *minimum* of n independent exponential random variables has also an exponential distribution with a parameter that is the sum of n parameters. Formally, if X_i are independent exponentially distributed random variables with parameter λ_i, $i = 1,...,n$, then $Y = \min\{X_1,...,X_n\}$ is also exponentially distributed with parameter $\sum_{i=1}^{n} \lambda_i$. Moreover, the probability that this minimum is attained by X_i is $\lambda_i \Big/ \sum_{j=1}^{n} \lambda_j$. This situation is called a *race of exponentials* and plays an important role in the theory of reliability and in combat modeling.

Normal Distribution

The occurrence of many real-world phenomena can be conveniently explained by the *normal* distribution, also called the *Gaussian* distribution. In general, the normal distribution can be said to apply to many situations involving measurements the frequency of whose values fall within symmetric patterns about a central value, with increasing frequency as one tends toward the central value. In particular, the sum of several random variables, of any distribution, tends to have this property, a fact which is formalized in the central limit theorem. For example, consider the values which might be obtained when taking any of the following measurements: (a) the heights of all men, age 21 or older, in a large city; (b) the grades received on the graduate record examination by all college seniors throughout the country; (c) the circumferences of all the full-grown trees in a large forest; and (d) the thicknesses of all the metal discs produced by an automatic machine in a 24-h period. The normal distribution has two parameters: the mean μ and the standard deviation σ. The pdf of the normal distribution, which has a distinctive bell shape, is

$$f(x) = \frac{1}{\sigma\sqrt{2\pi}} \exp\left(-\frac{1}{2}\left(\frac{x-\mu}{\sigma}\right)^2\right), \quad -\infty < x < \infty. \tag{A.19}$$

There is no closed form for the cdf. A normal distribution with parameters μ and σ is denoted by $N(\mu, \sigma)$.

Standard Normal Distribution

Any normal distribution can be *standardized* in order to use tabulated integrals for the special case where $\mu = 0$ and $\sigma = 1$. The standardized normal distribution is denoted by $N(0,1)$. It is common to denote the standard normal random variable by *Z*. The pdf of *Z* is

$$f(z) = \frac{1}{\sqrt{2\pi}} \exp\left(-\frac{z^2}{2}\right), \quad -\infty < z < \infty. \qquad \text{(A.20)}$$

The cdf of the standard normal distribution is denoted by $\Phi(z) = \int_{-\infty}^{z} \frac{1}{\sqrt{2\pi}} \exp\left(-\frac{u^2}{2}\right) du$ and is widely tabulated. In Excel™, the function is called NORMSDIST(). The transformation from a general normal distribution to a standard one is straightforward: $F(x) = \Phi\left(\frac{x-\mu}{\sigma}\right)$.

Example 19: The daily number of mortar shells launched toward the headquarters compound is normally distributed with mean 20 and standard deviation 4. The probability that on any given day a mortar attack comprises more than 23 shells is $P(X \geq 23) = 1 - P(X \leq 23) = 1 - \Phi\left(\frac{23-20}{4}\right) = 1 - \Phi\left(\frac{3}{4}\right) = 1 - 0.77 = 0.23.$

Bivariate Normal Distribution

A random variable may have more than one dimension. For example, the point of impact of a bomb relative to a target positioned at the coordinates (μ_X, μ_Y) is a bivariate random variable (*X*, *Y*), where *X* is the latitudinal error and *Y* is the longitudinal error with respect to the target. The latitudinal error can be negative (deflection to the left) or positive (deflection to the right) and the same is true for the longitudinal error where "short" is in the negative direction with respect to the target and "long" is in the positive direction. If the error in each dimension is normally distributed with means 0 and standard deviations σ_X and σ_Y for *X* and *Y*, respectively, and if the correlation between the two errors is ρ, then the pdf of the point of impact is, letting $K = \frac{1}{2\pi \sigma_X \sigma_Y \sqrt{1-\rho^2}}$

$$f(x,y) = K \exp\left(-\frac{1}{2(1-\rho^2)}\left[\left(\frac{x-\mu_X}{\sigma_X}\right)^2 - 2\rho\left(\frac{x-\mu_X}{\sigma_X}\right)\left(\frac{y-\mu_y}{\sigma_y}\right) + \left(\frac{y-\mu_Y}{\sigma_Y}\right)^2\right]\right).$$

(A.21)

If the two random variables are independent then $\rho = 0$, and

$$f(x,y) = \frac{1}{2\pi\sigma_X\sigma_Y}\exp\left(-\frac{1}{2}\left[\left(\frac{x-\mu_X}{\sigma_X}\right)^2 + \left(\frac{y-\mu_Y}{\sigma_Y}\right)^2\right]\right). \tag{A.22}$$

An important result in this context is as follows: if the point of impact has bivariate normal distribution with means (μ_X, μ_Y) and standard deviation (σ_X, σ_Y) and the two errors are independent, then the probability that the bomb lands within the ellipse centered at (μ_X, μ_Y) with latitudinal semi-axis length $r\sigma_X$ and longitudinal semi-axis length $r\sigma_Y$ is $1-\exp\left(\frac{-r^2}{2}\right)$. In particular, if $\sigma_X = \sigma_Y = \sigma$ then the pdf of the (Euclidean) distance of the point of impact from the target located at (μ_X, μ_Y) is $F(r) = 1-\exp\left(\frac{-r^2}{2\sigma^2}\right)$. $F(r)$ is called the *Rayleigh distribution.*

A.9 Stochastic Processes

A *stochastic process* (sometimes called *random* process) is an indexed collection or sequence of random variables. A stochastic process is denoted as $\{X(t), t \in T\}$ where T is called an index set. The most natural way to view T as *time*, but T could also represent other index sets such as spatial coordinates. The index set can be discrete, i.e., $T = \{0,1,2...\}$ or continuous, $T = [0,\infty)$. The *state space* of a stochastic process is the set of all possible values of $\{X(t), t \in T\}$.

Example 20: The number of aircraft in a squadron that are mission ready on any given day is a stochastic process. The index set is discrete (days) and the state space is $\{0,1,\ldots,n\}$, where n is the number of aircraft in the squadron.

We will focus on a special type of stochastic processes, those which possess the *Markov* property. But before we define the Markov property and present its applications, we first present a simple and widely used special case, the *Poisson process*.

A.10 Poisson Process

The Poisson process counts events occurring in time and is characterized by a parameter $\lambda > 0$, which is the rate of arrival of events (see *Poisson distribution* in Section A.7). The random variable in that process, usually denoted by $N(t)$, counts the number of arrivals up to time t. Note that, as a function of time, $N(t)$ is an integer monotone non-decreasing step function; it has a jump whenever an event occurs. There are three assumptions that underlie the Poisson process: (1) the arrival of two or more events simultaneously is unlikely, there is always a time interval $\Delta t > 0$ between two consecutive events; (2) for any sequence of points in time

$0, t_1, ..., t_n$ the process increments $\underbrace{N(t_1) - N(0)}_{\substack{\text{Number of arrivals} \\ \text{in the period } (0,t_1]}}, \underbrace{N(t_2) - N(t_1)}_{\substack{\text{Number of arrivals} \\ \text{in the period } (t_1,t_2]}}, ..., N(t_n) - N(t_{n-1})$ are independent random variables, that is, the *increments* of the process are independent random variables; and (3) the probability distribution of $N(t+s) - N(t)$ depends only on *s* (not on *t*), that is, the increments are *stationary* random variables. These three assumptions determine the probability distribution of $N(t)$:

$$P(N(t) = n) = \frac{(\lambda t)^n e^{-\lambda t}}{n!}. \tag{A.23}$$

The *inter-arrival* times in the Poisson process are exponentially distributed random variables, each with pdf $f(t) = \lambda e^{-\lambda t}$.

Suppose that each event that arrives in a Poisson process is one of *k* types. The probability that an event is of type i, $i = 1, ..., k$, is q_i. Then the arrival process of events of type *i* is also a Poisson process with arrival rate $q_i \lambda$. This property is sometimes called *thinning* of a Poisson process.

Example 21: The arrival of soldiers to the base health clinic follows a Poisson distribution with rate of 10 patients an hour. Thirty percent of the patients are female. The probability that during a 2-h period at most five females will show up at the clinic is

$$P(N(2) \le 5) = \frac{e^{-(.3)(10)(2)}}{0!} + \frac{(.3)(10)(2)e^{-(.3)(10)(2)}}{1!} + \cdots + \frac{((.3)(10)(2))^5 e^{-(.3)(10)(2)}}{5!} = 0.45.$$

The Poisson process described above is *homogeneous* in the sense that the arrival rate λ remains constant throughout the process. It does not depend on the specific time at which this rate is observed. A more general process, called *non-homogeneous Poisson process*, relaxes this assumption; the arrival rate $\lambda(t)$ is a non-negative function of time. In that case (A.21) becomes

$$P(N(t) = n) = \frac{\left(\int_0^t \lambda(x)dx\right)^n e^{-\int_0^t \lambda(x)dx}}{n!}. \tag{A.24}$$

A.11 Discrete-Time Markov Chain

Probably one of the most applied stochastic processes in practice is *Markov chain*. A discrete-time Markov chain is defined by the *states* of the stochastic process, the discrete *steps* or *stages* in which the states of the process are observed, the *initial probabilities* of the states, and the *transition probabilities* that represent the likelihood of moving from one state to another in one step. Formally, a stochastic process $(X_n, n = 0, 1, 2, ...\}$ is said to be a Markov chain if for all states $i_0, i_1, ..., i_n$ and *j*, the probability that the next state is *j*, given a complete history of previous states, depends only on the most recent state:

$$P\{\underbrace{X_{n+1}=j}_{future}\,|\,\underbrace{X_0=i_0,X_1=i_1,\dots,X_{n-1}=i_{n-1}}_{past},\underbrace{X_n=i_n}_{present}\}=P\{\underbrace{X_{n+1}=j}_{future}\,|\,\underbrace{X_n=i_n}_{present}\}. \quad \text{(A.25)}$$

In other words, the future is *conditionally* independent of the past, given the present. A Markov chain is said to have *stationary* transition probabilities if $p(X_{n+1}=j|X_n=i)=p_{ij}$ – independent of the time step n. We will consider only stationary Markov chains. Let P denote the transition matrix of a Markov chain with m states as follows:

$$P=\begin{bmatrix} p_{11} & p_{12} & \cdot & p_{1m} \\ \cdot & \cdot & \cdot & \cdot \\ \cdot & \cdot & \cdot & \cdot \\ p_{m1} & p_{m2} & \cdot & p_{mm} \end{bmatrix}. \quad \text{(A.26)}$$

The three properties of P are (1) P is a square $m\times m$ matrix; (2) the sum of each row is 1; and (3) all entries in P are numbers between 0 and 1.

Given a vector $\pi(0)=(\pi_1(0),\dots,\pi_m(0))$, where $\pi_i(0)=P(\text{the process is initially at state } i)$, the probabilities of the various states at step 1 are given by $\pi(1)=\pi(0)P$, that is, $\pi_i(1)=\sum_{j=1}^{m}\pi_j(0)p_{ji}$. Similarly, the probabilities of the states at the nth step are given by $\pi(n)=\pi(n-1)P=\pi(n-2)P^2=\cdots=\pi(0)P^n$. It follows that an entry $p_{ij}(n)$ in the matrix P^n is the probability that the system moves from state i to state j in n steps.

Example 22: A system can be in one of three states: *good* (G), *bad* (B), and *down* (D). Let X_n denote the state of the system at the nth inspection. Thus, the *state space* is $\{G, B, D\}$ and the *steps* are inspection instances. Initially the system is in state G and the transition matrix is

$$P=\begin{matrix} G \\ B \\ D \end{matrix}\begin{pmatrix} 0.8 & 0.1 & 0.1 \\ 0.2 & 0.2 & 0.6 \\ 0.4 & 0.5 & 0.1 \end{pmatrix}.$$

For example, if the system is in good condition, then the probability it is going to be in bad condition at the time of the next inspection is 0.1. Similarly, if the system is currently down, the probability that in the next inspection it is fully repaired and is back in good condition is 0.4. With probability 0.1 it remains down. Since it is given that initially the system is in good condition it follows that

$\pi(0) = (1,0,0)$. The probability vector of the state of the system at the second inspection is

$$\pi(2) = (1,0,0)\begin{pmatrix} 0.8 & 0.1 & 0.1 \\ 0.2 & 0.2 & 0.6 \\ 0.4 & 0.5 & 0.1 \end{pmatrix}^2 = (0.7, 0.15, 0.15).$$

Thus, the probability that the system is in good condition two steps (inspections) from now is 0.7, while the probability it is down is 0.15.

A.12 Continuous-Time Markov Chain

Recall that in a discrete-time Markov chain the process is observed at fixed points in time called steps or stages. But what happens in between two consecutive steps? What would happen if instead of observing the system every hour we will observe it every half hour? 10 min? 1 s?

As the name implies, in a continuous time Markov chain (CTMC) the process is observed continuously. Instead of considering the discrete-time stochastic process $\{X_n, n = 0,1,2,...\}$ we consider now its continuous counterpart, that is, $\{X_t, t \geq 0\}$. In that case the Markov property becomes $P\{X(t+s) = j \mid \text{path of the process in } [0,t),\ X(t) = i\} = P\{X(t+s) = j \mid X(t) = i\}$. The process has stationary transition probabilities if $P\{X(t+s) = j \mid X(t) = i\} = p_{ij}(s)$ is independent of *t*.

We have already seen one CTMC – the Poisson process *N*(*t*). Here we have $P\{N(t+s) = j \mid N(t) = i\} = P(N(s) = j - i)$, as given by (A.23).

Any CTMC comprises a series of jumps that occur at certain points in time. Between any two jumps, say from state *i* to state *j*, the process remains unchanged at state *i*. The length of time the process spends at any given state *i* is called the *sojourn time* in state *i*. The Markov property implies that the sojourn times are independent random variables having exponential distributions. The mean sojourn time generally depends on the state the process is in, although this is not the case for a Poisson process where we already know (Section A.10) that the mean sojourn time is the same for all states and is equal to $1/\lambda$. The successive states visited in a CTMC form a discrete-time (event-driven) Markov chain where a *step* represents a transition from one state to another. Thus, a CTMC is completely defined by a state transition matrix $K = (k_{ij})$ that specifies the transition probabilities from one state to another, and the mean sojourn time $1/v_i$ in each state *i*.

Example 23: In a Poisson process $v_i = \lambda$ for all states *i*, and $k_{ij} = 1$ if $j = i + 1$, and $k_{ij} = 0$ otherwise.

Example 24: A machine is subject to failure and repair. Repair times are independent identically distributed exponential random variables with mean $1/\mu$. Successive up times are independent identically distributed exponential random variables with mean $1/\lambda$. Here $v_{up} = \lambda$ and $v_{down} = \mu$. The state transition matrix is 2×2 where $k_{01} = k_{10} = 1$ and $k_{00} = k_{11} = 0$.

Example 25: Consider a duel between two combatants *B* and *R*. The time it takes *B* to kill *R* is exponentially distributed random variable with parameter *b*. The time it takes *R* to kill *B* is exponentially distributed random variable with parameter *r*. The states of this process are (1,1) (combat in progress), (1,0) (*B* wins), and (0,1) (*R* wins). Here $v_{(1,1)} = b + r$, $v_{(1,0)} = v_{(0,1)} = 0$, $k_{(1,1),(1,0)} = b/(b+r)$, $k_{(1,1),(0,1)} = r/(b+r)$, $k_{(0,1),(0,1)} = k_{(1,0),(1,0)} = 1$, and $k_{ij} = 0$ for all other ij (see *race of exponentials* in the *exponential distribution* subsection of Section A.8).

An important class of CTMCs are *absorbing chains* where some states are absorbing; once the process enters one of these states it stays there forever. The other, non-absorbing, states are called *transient* states. The Markov transition matrix *K* of such a CTMC is of the form

$$K = \begin{bmatrix} I & 0 \\ R & Q \end{bmatrix}, \tag{A.27}$$

where I is a unit matrix that corresponds to the absorbing states, Q is a square submatrix containing the transition probabilities among transient states, and R is a submatrix that represents transitions from a transient state to an absorbing state. Such a matrix has some very useful properties. First note that in the long run the process will end up in one of the absorbing states. The limiting probabilities of the various absorbing states are given in the matrix $(I-Q)^{-1}R$. The rows of this matrix correspond to the transient states and its columns represent the absorbing states, thus the i, j entry of $(I-Q)^{-1}R$ is the probability that the Markov process ends up in (absorbing) state *j* given it starts in (transient) state *i*. Second, the expected number of visits to a certain transient state *l*, given that the process starts in (transient) state k is the k,l entry of the matrix $(I-Q)^{-1}$.

Example 26: Consider Example 25. The duel between *B* and *R* is represented by the matrix

$$\begin{bmatrix} 1 & 0 & 0 \\ 0 & 1 & 0 \\ \frac{b}{b+r} & \frac{r}{b+r} & 0 \end{bmatrix}.$$

Here, $i=1$ is the absorbing state (1,0), $i=2$ is the absorbing state (0,1), and $i=3$ is the (only) transient state (1,1). Also $R=\left[\frac{b}{b+r},\frac{r}{b+r}\right]$ and $Q=0$. The limiting probabilities are $(1-0)^{-1}\left[\frac{b}{b+r},\frac{r}{b+r}\right]=\left[\frac{b}{b+r},\frac{r}{b+r}\right]$ and the expected number of visits to a transient state is 1 – the initial state of the duel.

Appendix B
Optimization

The main reason for modeling combat is to improve decision making, whether the decisions concern procurement, operations, tactics, or logistics. In certain limited circumstances, it is possible to find the *best possible* decision. Methods for doing this with the aid of a computer are the subject of this section. Computations will be illustrated using the Excel™ workbook *AppendixB.xls*.

This appendix is a brief introduction to a topic about which there is much more to be said. For further details and more examples, see Winston (1994) or some other book that introduces operations research.

B.1 Introduction

John von Neumann and Morgenstern (1944) prove that "rational" decision makers will make decisions that maximize the decision maker's expected "utility", a scalar measure of desirability. The widespread use of computers to help make decisions through optimization is based on this foundation. In practice we may minimize, rather than maximize, since sometimes it is more convenient to deal with disutility (cost), rather than utility.

If $\mathbf{x}$ is the decision to be made and $U(\mathbf{x})$ is the utility of the decision, the associated mathematical problem is to maximize $U(\mathbf{x})$ subject to $\mathbf{x}$ being feasible. The decision symbol is in bold type to emphasize that it may be a vector rather than a simple number. The utility function itself must be a scalar. If the number of possible decisions is not too large, it may be possible to simply examine them all and select the one with the largest utility. In fact, the "sort" function in any spreadsheet might be considered the most elementary of its optimization functions, since sorting on utility will leave the best decision exposed. This might be called brute force optimization – simply list all the possibilities and choose the best.

However, there are many problems where it is not attractive or even impossible to explicitly list all possible decisions. In Excel™, this is handled by Solver, a tool that brings up a dialog box where the user is invited to select one utility cell to optimize (you must be explicit about whether you want to maximize or minimize), and one or more "adjustable" cells that Solver will adjust to optimize utility. These adjustable cells are Solver's name for $\mathbf{x}$, the decision, and may be subjected to various feasibility constraints. If you do not see Solver on the Tools menu, check the box on the list of Add-ins so that you will have it available.

Example 1: Consider the maximization of the function $f(x) = \sin(x)/x$, subject to the constraint that x must be at least 3. On sheet "Hill Climb" of *AppendixB.xls*, you can see a graph of that function, from which it is obvious that the maximum is at approximately $x = 8$ (better values are available for small x, but x is required to be at least 3). Solver is set up to find the maximum of that function, which is in the pink cell (F2), by adjusting the green cell (x, or G2). If you put any number near 8 in the green cell and run Solver, it will adjust the green cell to 7.7, which is optimal.

Note that Solver is careful in describing its answer:

Solver has found a solution. All constraints and necessary conditions are satisfied.

That is not quite the same thing as saying that the solution is guaranteed to be optimal. In fact, if you start Solver with a number like 4 in the green cell, it will make the same speech after proposing that the solution is 3. Solver is basically hill climbing, by which we mean that it senses the slope of the function, and continually moves uphill until it gets to the top (which it recognizes because the function is flat there) or can go no further on account of a constraint. While this will lead to an optimum in the sense that small changes to x will produce no improvement to $f(x)$, the final point may still not be optimal in a global sense.

Of course, we would prefer it if computed solutions could actually be guaranteed to be globally optimal. The principal class of problems where this is actually true is considered next.

B.2 Linear Programs

In a linear program (LP), $\mathbf{x}=(x_1,\ldots,x_n)$ is a vector with n components, the objective function is $U(\mathbf{x})=\sum_{i=1}^{n} c_i x_i$, where $\mathbf{c}$ is an arbitrary vector of constants, and the constraints on $\mathbf{x}$ take the form that other linear combinations cannot exceed given limits. If there are m such constraints, they can all be compactly represented by letting $\mathbf{x}$ be a column vector and requiring $\mathbf{Ax} \le \mathbf{b}$, where $\mathbf{A}$ is a $m \times n$ matrix and $\mathbf{b}$ is a constant column vector with m components. The objective can be either maximized or minimized, and the constraints can also include equalities or $\ge$ inequalities.

Linear programs with millions of variables and constraints are solvable with modern computer codes, and the results are (or should be) guaranteed to be optimal, with three caveats. One caveat is that there may be no feasible solution, in which case there is of course no optimal solution. Another is that the constraints may not limit the solution, so that $U(\mathbf{x})$ is unbounded. For example, the problem of maximizing x subject to $x \ge 0$ is a linear program with an unbounded solution. Analysts do not usually formulate unbounded or infeasible LPs deliberately, but computer codes like Solver must be prepared for anything. The last caveat is that the LP may simply be too large for the computer code to handle. Subject to these three caveats, the optimal solution of any LP is computable.

Solver in the 2003 version of Excel™ can handle only 200 adjustable cells ($n \le 200$), but that is sufficient for tutorial purposes, as well as for many applications.

Example 2: This is a surveillance problem where there are three platform types available for maintaining surveillance over four areas. The number of platforms of each type i is an input b_i. The platforms have different efficiencies in the different areas. Aircraft, for example, are most efficient at providing surveillance in areas that are near their bases. One platform of type i, if assigned to area j, will find any target in that area with rate a_{ij}. The 12 inputs a_{ij} reflect the relative efficiencies of the three platform types in the four areas. The quantification of these inputs might involve the ideas of Chapter 7, especially Section 7.3.2, but here we take them as given. Let x_{ij} be the number of platforms of type i assigned to patrol in area j, and let $y_j \equiv \sum_i a_{ij} x_{ij}$ be the total detection rate in area j. This is a linear equation that relates the decision variables to a measure of effectiveness in each area. The object is to assign platforms so that the worst-case total detection rate is as large as possible. This is most easily accomplished by introducing one more variable, the objective v, while requiring that y_j be at least v in all four areas. Once the adjustable cells $(\mathbf{x},v)$ are optimized, the detection rate will be at least v in each of the four areas. Another way of putting this is that the mean time to detection will be at most $1/v$. In summary, the LP is to maximize v, subject to the constraints $\sum_i a_{ij} x_{ij} - v \geq 0$ for all j, $\sum_j x_{ij} \leq b_i$ for all i, and $x_{ij} \geq 0$ for all i and j. No explicit reference to y_j is necessary, so the LP has a total of 13 variables ($n = 13$). The constraints $x_{ij} \geq 0$ are necessary; otherwise, Solver will try to make some of them negative. Counting these 12 constraints, there are a total of 19. See sheet "MaxRate" of *AppendixB.xls* for a solution. The solution will be globally optimal, regardless of the starting point for Solver, because the optimization problem is an LP. Suggested exercises based on this example are

1) Examine the spreadsheet to see how the problem is set up for Solver. Solver does not permit linear expressions in its dialog box, so all such computations must be in the spreadsheet. Solver does not permit variables on the right-hand side of inequalities, so all variables must appear on the left, as in the formulation above. Solver does permit arrays on the left-hand side, so only one line in the dialog box is needed for all of the constraints $x_{ij} \geq 0$.

2) Change the input data **a** and/or **b** and run Solver to see the impact on **x** and v.

3) Delete the constraints $x_{ij} \geq 0$ to see what happens if you make a mistake like that.

B.3 Integer Constraints

The solution of Example 2 usually involves fractional values for some of the x_{ij} variables. This might or might not be a problem. It might be possible to achieve the equivalent of assigning (say) 2.5 aircraft to a particular area by assigning 2 on some days and 3 on others. In that case the solution of the LP is exactly what is needed. If this kind of time averaging is not possible, however, some variables will have to be restricted to integer values. The result is known as an integer program if all of the variables are so restricted, or as a mixed integer program (MIP) in the more common case where only some of them are. In the case of the surveillance example, we would restrict **x** to be integer, but not v, so we would have a MIP. In either case, the resulting optimization problem is of a fundamentally more difficult type than is an ordinary LP.

The added computational difficulty need not necessarily concern the user. In using Solver, it is simply a matter of adding a constraint that certain variables must be integers. In the dialog box, add a constraint where all of the subject variables are named, and select "integer" from the dropdown relationship menu. The main import of the computational difficulty is that the user should have a heightened sense of apprehension about what will happen when the "solve" button is pushed. There are examples where no solution at all is returned within reasonable time, and also examples where an incorrect solution is returned.

Example 3: The knapsack problem is a famous integer programming problem where a backpacker can only carry **b** pounds, and must decide which items to put in his knapsack. Each item has a weight and a value, and any number of items of each type can be carried as long as the knapsack does not become heavier than b. The value of a collection of items is by assumption additive (which means that the naming of this kind of problem was unfortunate, since the value of three sleeping bags is surely not three times the value of one). The knapsack problem is trivial without the integer restriction – the solution is to carry as many as possible of the item with the greatest value/weight ratio. The integer restriction puts the problem into a class that is known to be fundamentally difficult. As of this writing, Solver gets the wrong solution to at least one knapsack problem, the one shown on sheet "Knapsack" of *AppendixB.xls*. The optimal solution is known to be $\mathbf{x} = (32,2,1,0,0,0,0)$, with a total value of 19,979. The Solver that comes with Excel™ 2003 gets a score of only 19,972, even if the optimal solution is provided as a starting point.

Example 3 is not meant to convince you that using Solver is a bad idea; 19,972 is not a bad approximation to 19,979, after all, and Solver usually gets exactly the right answer. Nonetheless, bear in mind that, just as there are some remarkably complex looking problems are actually simple (any LP), there are also some remarkably simple sounding problems that are actually complex enough to justify skepticism about "optimal" solutions.

Integer variables are sometimes restricted to take on only two values, 0 and 1, in which case they are called "binary" variables. Binary variables can be used to include certain logical relationships in a problem that would otherwise be an LP. Suppose that x represents the level of production of some commodity, and that the objective function includes a term $-kx$ because each unit of x costs k to produce. The trouble is that a factory is required if the level of production is to be positive, but currently there is no factory. Introduce a binary variable z that indicates whether a factory is to be built, together with the term $-Kz$ in the objective function, where K is the cost of a factory. The logical idea that nothing can be produced without a factory can be included by adding the constraint $x \leq Mz$, where M is a factory's production capability. If z is 0, then x cannot be positive. If z is 1, then x can be as large as M. This method of including the idea that one system is required for the employment of another has wide applicability. It is employed in Section 10.2.2, for example, to enforce the idea that UAVs cannot fly without a ground control unit. There are special algorithmic techniques for handling binary variables, so it is wise to identify them as such (not as integer variables) when using most optimization packages. In Solver, the same pulldown list that includes "int" also has "binary" on it. Choose the latter, when appropriate.

B.4 Nonlinear Programs

If any of the functions involved in expressing the decision problem are nonlinear functions of $\mathbf{x}$, the result is called a nonlinear program (NLP). Example 1 is a nonlinear program that illustrates the tendency of nonlinear solvers to find a locally optimal solution, but not necessarily a globally optimal one. Another illuminating exercise would be to try maximizing x^2 with Solver, starting from an initial guess of 0. Nothing happens in the Solver that comes with Excel™ 2003, the difficulty being that the function is flat at the origin (with any other starting point, Solver will correctly report that the function is unbounded).

Again, these cautionary tales are not meant to inhibit your use of Solver or other optimization software, but only to instill the habit of skepticism. Nonlinear problems are solved in several places in this book, with results being presented as "optimal" after only reasonableness tests on the part of the authors. Most of the time, in most reasonably posed problems, Solver will get the optimal solution. Besides, employing Solver is fail-safe – if you do not like Solver's solution, you can always reject it in favor of the starting point you provided in the first place.

It is worth some effort to try to express an NLP as an LP, if possible. An example of that might be Example 2, where y_j was defined to be the total search rate in area j. We might maximize the function $\min(y_1,y_2,y_3,y_4)$, the smallest of the four search rates. This would cause us to save one variable (v) at the cost of making the objective function nonlinear, since min() is a nonlinear function. This is a no-brainer – there is great value in preserving the LP, and only a small sacrifice in introducing one more variable. The linear formulation given in Example 2 is best. Another example is in Section 10.2.2, where a nonlinear problem is converted into a linear problem by taking logarithms.

B.5 Dynamic Programming (DP)

Solver will not solve dynamic programs, nor will most other optimization packages. The reason is that DP problems do not take a standard form, but instead exploit a principle called the "principle of optimality." This principle implies that the solution of a large problem may automatically imply the solution of a host of smaller problems. For example, suppose that the shortest route from New York to Los Angeles goes through Pittsburgh. Then it follows that the part of it going to Pittsburgh is also the shortest route from New York to Pittsburgh. This comparatively innocent observation actually has considerable power, to the point where it underlies not only shortest route computations but also computations in similar circumstances where a convenient "state" can be identified. The state of a shortest route problem is the city where one currently is. The state of the ABM problems considered in Chapter 4 is the number of ABMs remaining to allocate to whatever number of attackers survive the previous allocations.

The basic DP method is to embed the given problem in a class of smaller problems interconnected by a recursive equation. At the end of computations, one has typically solved all smaller problems, as well as the given problem. The solution method does not take well to direct implementation in a spreadsheet, but a programming language such as VBA (used in Chapter 4) is a sufficient extension. The inventor of DP is Richard Bellman, whose book with Dreyfus (1962) is still worth reading. Alternatively, see Winston (1994).

Appendix C
Monte Carlo Simulation

Monte Carlo simulation is a method for investigating stochastic models that involves the use of random numbers, typically as generated by a computer using a random number generator. These random numbers can be used to imitate the kind of unpredictability sometimes exhibited by the real world. Through repetition, random variables of interest can often be quantified, even when direct probabilistic analysis fails. The technique is an old one, but did not become practical for accurate computations until the advent of the digital computer. It was used by E. L. de Forest in the Manhattan Project of World War II, at which time Stanley Ulam and John von Neumann named the technique.

C.1 The General Idea

Suppose that random variable X is of interest, but that the probability distribution of X is unknown. One method of describing X is to repeatedly perform the fundamental experiment on which X is based, observe X_i in the ith independent iteration, and then use statistical methods to infer something about X from the random sample $X_1, \ldots, X_n$. If n is large, we can generally expect our inferences to be accurate. The random sample might be obtained from observations about the real world, but in Monte Carlo simulation it is generated artificially, nearly always using a computer.

It may seem remarkable that a random sample $X_1, \ldots, X_n$ could be generated under the premise that the probability distribution of X is unknown, but exactly that situation is remarkably common. For example, let X be the sum of two six-sided dice, as in the game of Craps, and consider the event that X is 11. What is $P(X = 11)$? Since X is not a standard kind of random variable, we cannot look the answer up in a table. However, samples of X are easy to generate by simply rolling two dice and adding up the spots. "Rolling a die" is easy on a computer, so we can perform the experiment (say) 10,000 times and then estimate $P(X=11)$ to be the fraction of times when $X = 11$. The experimental fraction will not be exactly correct, but it will be close to the exact value of 1/18.

If you already know that the exact answer is 1/18, you might be wondering why anyone would be foolish enough to go to all that trouble to be slightly wrong about the chances of getting 11 in Craps. People who are skilled at probability calculations often have that reaction to Monte Carlo simulation. However, while it is certainly true that there is no point in simulating something to which you already know the answer, it is also true that there is a distressing abundance of problems for which exact answers are lacking, even for skilled probabilists. Let X be the sum of 10 dice. Do you know $P(30 \leq X \leq 50)$? To calculate that number exactly, you would first calculate $P(X = 30)$, and then add that to 20 other similar numbers. To calculate $P(X=30)$, you would need to count the number of ways in which 10 six-sided dice could sum to 30. All of this could certainly be done, but the process is time consuming and error prone. Why not just roll 10 dice to determine *X*, and repeat the experiment often enough to be confident in the experimental fraction? The replications can be performed very rapidly on a computer.

For a more important example, consider nuclear fission. When a uranium atom releases a neutron, the neutron will travel in a uniformly random direction until it encounters something. It might encounter another uranium atom, thus generating more neutrons, or it might be absorbed by some other kind of atom. The process might generate so many neutrons that it would become unstable, or it might not. To estimate the chances of instability, the Manhattan scientists employed tables of random numbers to determine the details of individual interactions. Modern computers simply make the Monte Carlo process more efficient.

C.2 Random Number Generation

Computers are designed to be absolutely predictable in their computations, so finding a satisfactory way to generate "random" numbers is a challenging task. One method is to consult the system clock at the exact time when the number is needed, basing the result on the low-level digits. This method is seldom used because it is not reproducible – even though we wish to generate "random" numbers, we also wish to be able to reproduce our experiments for debugging purposes. The usual compromise is to generate "pseudorandom" numbers in a sequence x_1, x_2, …, where the generation of x_{n+1} depends on x_n. For example, one famous method is to let x_{n+1} be the remainder after dividing ax_n by m, where a and m are constant positive integers. If m is a prime number, then x_{n+1} will always be an integer larger than 0 and smaller than m. If a and m are carefully chosen, the sequence will pass most statistical tests for being a sequence of independent random integers uniformly distributed over that interval, but will still be reproducible if necessary because the entire sequence depends on the "seed" x_1. One particularly apt choice for 32-bit computers is $a = 7^5$ and $m = 2^{32} - 1$, which Park and Miller (1988) refer to as the "minimal standard."

Most programming languages contain a function for generating a sequence of random numbers uniformly distributed over the interval (0,1). One way of doing this would be to divide the minimal standard sequence by m, but the programmer need not be concerned with how the sequence is actually generated – whenever a random number is needed, it suffices to call the function. In Excel™, this function is called RAND(). RAND() is a "volatile" function; if you write "=RAND()" in two different cells, the numbers displayed will be different in the two cells, and furthermore will change every time you make any change in the spreadsheet. Furthermore the effects of "=2*RAND()" are different from "=RAND()+RAND()," since the second expression calls the RAND() function twice while the first calls it only once.

Given the availability of random numbers uniformly distributed in the interval (0,1), it turns out not to be difficult to generate random numbers having other distributions. The theoretical reason is that, if *F*() is the cumulative distribution function of random variable *X*, then *F*(*X*) is a uniform random variable. Since we can generate uniform random variables, we can also generate random variables from *F*() by employing its inverse function. In Excel™, for example, the expression "=NORMINV(RAND(), A1, A2)" will produce a normal random number with mean A1 and standard deviation A2. There are a variety of commercial "addins" that provide functions for sampling from practically any standard distribution.

Once we are able to generate random numbers with arbitrary distributions, we can employ them to simulate a variety of interesting problems that would otherwise be difficult to analyze.

C.3 Statistical Issues and SimSheet.xls

Monte Carlo simulation works by repeating the same experiment over and over, collecting and eventually displaying statistics about the outcome. Suppose that the outcome is a number *Y*, a random variable that summarizes the experiment. Natural statistics to report would be the sample mean, the standard error of estimating the mean, and perhaps a histogram showing the sample distribution. *SimSheet.xls* automates this process. That workbook contains an input cell A1 that refers to whatever cell defines *Y*, and another input cell that specifies the number of replications desired. After adjusting these two cells, the user presses the "Simulate" command button to repeat the experiment the desired number of times. Outputs include those suggested above. *SimSheet.xls* can be employed with any other workbook, as long as both workbooks are open at the same time. On SimSheet, type "=" in cell A1. Then click the cell in the other workbook that you want to simulate, and press the green checkmark.

The number of replications can be selected to make the standard error "sufficiently small," which of course depends on the application. A few thousand replications usually suffice. Confidence intervals can be constructed by the usual statistical techniques. A confidence interval for *E*(*Y*), for example, can be based on the sample mean and standard error reported by *SimSheet.xls*.

C.4 Examples and Exercises

Example 1: Consider the probability of getting 11 in Craps. A Monte Carlo simulation is on sheet "Craps" of *AppendixC.xls*. A die roll is put into each of two cells, and the sum is put into a third. Using *SimSheet.xls*, select (say) 10,000 replications, press the simulate button, and examine the resulting histogram to determine the fraction of rolls that are actually 11. The answer is unpredictable because RAND() is a volatile function, but should be close to 1/18. Alternatively, add the formula "=IF(C2=11,1,0)" to cell D2, which will put a 1 or a 0 into cell D2 depending on whether the total is 11 or not. The average of this quantity, which is automatically reported by *SimSheet.xls* if you link it to cell D2, is the fraction of replications where an 11 appears.

Example 2: The Battle of 73 Easting was a battle in the first Gulf War where 23 red tanks were killed by 12 blue tanks, with no losses to blue. Particularly remarkable was that seven of these kills were made by a single blue tank. Or is it remarkable? We would like to know the probability that this might happen by chance, rather than through any special skill or circumstance affecting the record setting blue tank. To be precise, what is the probability P that the highest scoring blue tank would kill seven or more red tanks even if each red kill were made by a randomly selected blue tank? Calculation of P is difficult, but simulation is not. Sheet "73Easting" of *AppendixC.xls* does this. Each row of that sheet determines the fate of one red tank by using a single random number (using a new random number for every cell in the row would not give the desired result) to select its killer. The Sum() and Max() functions of Excel™ are then sufficient to determine the number of red tanks killed by the highest scoring blue tank, recorded in cell N25. Cell N25 can then be linked to *SimSheet.xls*. Probability P turns out to be approximately 0.02, so the chances of this lopsided tally happening "by chance" are small enough that some other explanation should be sought.

There are other examples in the text. Example 3 of Section 6.3 involves a Monte Carlo simulation of a Lanchester battle. *Chapter 8.xls* includes separate sheets for the same minefield planning problem solved analytically and by Monte Carlo simulation – see Section 8.3 for a detailed description.

C.5 Further Reading

The above brief introduction is sufficient for this book, but there is much more to be said about simulation. Books on operations research such as Taha (2007) or Winston (1994) often include a section on Monte Carlo simulation, including methods for generating random samples from various distributions. In addition to addins for Excel™, there are special-purpose simulation software packages available that aid visualization and the formulation of complicated problems. Swain (2005) is a survey of over 40 different packages available.

References

Ahuja, R. K.; Magnanti, T. L.; and Orlin, J. B. 1993. "Network Flows: Theory, Algorithms and Applications", *Englewood Cliffs*, NJ: Prentice Hall. [9.2.1]

Ahuja, R.; Kumar, A.; Jha, K.; and Orlin, J. 2003. "Exact and Heuristic Methods for the Weapon Target Assignment Problem", MIT Sloan School of Management Working Paper 4464-03, Cambridge. [2.4.3]

Alighanbari, M. and How, J.P. 2008. "A Robust Approach to the UAV Task Assignment Problem", *International Journal of Robust Nonlinear Control*, vol. 18, pp. 118–134. [9.4.1]

Anderson, L. 1989. "A Heterogeneous Shoot-Look-Shoot Attrition Process", paper P-2250, Institute for Defense Analysis, pp. 10–11. [3.4.2]

Anderson, R.M., and May, R.M. 1991. *Infectious Diseases of Humans*, Oxford University Press, Oxford. [10.3.2]

Aviv, Y. and Kress, M. 1997 "Evaluating the Effectiveness of Shoot-Look-Shoot Tactics in the Presence of Incomplete Damage Information", *Military Operations Research*, vol. 3, pp. 79–89. [3.5]

Barfoot, C. 1969. "The Lanchester Attrition-Rate Coefficient: Some Comments on Seth Bonder's Paper and a Suggested Alternate Method", *Operations Research*, vol. 17, pp. 888–894. [5.4.1]

Barr, D. 1974. "Strong Optimality of the Shoot-Adjust-Shoot Strategy", *Operations Research*, vol. 22, pp. 1252–1257. [3.3]

Beale, E. and Heselden, G. 1962. "An Approximate Method of Solving Blotto Games", *Naval Research Logistics*, vol. 9, pp. 65–79. [6.2.3]

Bellman, R. 1961. *Adaptive Control Processes: A Guided Tour*, Princeton University Press, Princeton. [3.4.2]

Bellman, R. and Dreyfus, S. 1962. *Applied Dynamic Programming*, Princeton University Press, Princeton. [Appendix B]

Benkoski, S.; Monticino, M.; Weisinger, J. 1991. "A Survey of the Search Theory Literature", *Naval Research Logistics*, vol. 38, pp. 469–494. [7.7]

Berkovitz, L. and Dresher, M. 1959. "A Game theory Analysis of Tactical Air War", *Operations Research*, vol. 7, pp. 599–620.[6.2.3, 6.3]

Bertsekas, D.; Homer, M.; Logan, D.; Patek, S.; and Sandell, N. 2000. "Missile Defense and Interceptor Allocation by Neuro-Dynamic Programming", *IEEE Transactions on Systems, Man and Cybernetics*, *Part A*, vol. 30, pp. 42–51. [4.3]

Bracken, J.; Falk, K.; and Karr, A. 1975. "Two Models for Allocation of Aircraft Sorties", *Operations Research*, vol. 23, pp. 979–995. [6.2.3]

Bracken, P. 1976. "Unintended Consequences of Strategic Gaming", Hudson Institute HI-2555-P, New York. [6.3]

Bracken, J.; Kress, M.; and Rosenthal, R. 1995 (eds), *Warfare Modeling*, Wiley, New York. [5.2]

Referring sections are shown in [ ]

Brackney, H. 1959. “The Dynamics of Military Combat”, *Operations Research*, vol. 7, pp. 30–44. [5.4.1]

Bressel, C. 1971. “Expected Target Damage for Pattern Firing”, *Operations Research*, vol. 19, pp. 655–667. [2.3.3]

Bronowitz, R., and Fennemore, C. 1975. “Mathematical Model Report for the Analytical Countered Minefield Planning Model (ACMPM)”, NSWC/DL TR-3359, Naval Surface Weapons Center. [8.5.1]

Brown, S. 1980. “Optimal Search for a Moving Target in Discrete Time and Space”, *Operations Research*, vol. 28, pp. 1275–1289. [7.6.2]

Brown, G.; Carlyle, M.; Diehl, D.; Kline, J.; and Wood, W. 2005. “A Two-Sided Optimization for Theater Ballistic Missile Defense”, *Operations Research*, vol. 53, pp. 263–275. [4.3, 6.2.4]

Brown, G., and Washburn, A. 2007. “The Fast Theater Model (FATHM)”, *Military Operations Research*, vol. 12, pp. 33–45. [5.6]

Bryan, K. 2006. “Algorithms for Decision Aid for Risk Evaluation (DARE) version 2.1”, NURC-FR-2006–002, NATO Undersea Research Centre. [8.4]

Burr, S.; Falk, J.; and Karr, A. 1985. “Integer Prim-Read Solutions to a Class of Target Defense Problems”, *Operations Research*, vol. 33, pp. 726–745. [4.2.3]

CAA. 1983. “ATCAL: An attrition Model using Calibrated Parameters”, The U.S. Army’s Center for Army Analysis, CAA-TP-83-3. [5.5]

CAA. 1998. “Kursk Operations Simulation and Validation Exercise – Phase II (KOSAVE II)”, The U.S. Army’s Center for Army Analysis, CAA-SR-98-7. [5.4.2, 5.5]

Cunningham, L. and Hynd, W. 1946. “Random Processes in Air Warfare”, *Journal of the Royal Statistical Society*, supplement, vol. 8, pp. 62–85. [2.5]

David, I. and Kress, M. 2005. “‘No Overlap No Gap’ and the Attack of a Linear Target by n Different Weapons”, *Journal of the Operational Research Society*, vol. 56, pp. 993–996. [2.5]

Davis, P. 1995. “Aggregation, Disaggregation, and 3:1 Rules in Ground Combat”, RAND report MR-638-AF/A/OSD, Santa Monica. [5.5]

Deitchman, S. 1962. “A Lanchester Model of Guerilla War”, *Operations Research*, vol. 10, pp. 818–827. [5.2]

Dell, R.; Eagle, J.; Martins, G.; and Santos, A. 1996. “Using Multiple Searches in Constrained-Path Moving-Target Search Problems”, *Naval Research Logistics*, vol. 43, pp. 463–480. [9.4.1]

DIA. 1974. “Mathematical Background and Programming Aids for the Physical Vulnerability System”, DI-550-27-74, Defense Intelligence Agency, Washington. [2.2.4]

DoD. 1998. “DoD Modeling and Simulation (M&S) Glossary”, DoD 5000.59-M January 1998. [1.2]

Eckler, A. and Burr, S. 1972. *Mathematical Models of Target Coverage and Missile Allocation*, Military Operations Research Society, Alexandria. [2.5, 4.3, 6.2.3]

Engel, J. 1954. "A Verification of Lanchester's Law", *Operations Research*, vol. 2, pp. 163–171. [5.4.2, 5.5, 6.2.3]

FAO. 2008. http://www.fas.org ,Federation of American Scientists, accessed November, 2008.

Fraser, D. 1951. "Generalized Hit Probabilities with a Gaussian target", *Annals of Mathematical Statistics*, vol. 22, pp. 248–255. [2.5]

Fraser, D. 1953. "Generalized Hit Probabilities with a Gaussian target II", *Annals of Mathematical Statistics*, vol. 24, pp. 288–294. [2.5]

Fricker, R. 1998. "Attrition Models of the Ardennes Campaign", *Naval Research Logistics*, vol. 45, pp. 1–22. [5.4.2]

Gal, S. 1980. *Search Games*, Academic Press, New York. [7.6.1]

Gilliland, D. 1962. "Integral of the Bivariate Normal Distribution over an Offset Circle", *Journal of the American Statistical Association*, vol. 57, pp. 758–768. [2.2.2]

Glazebrook, K. and Washburn, A. 2004. "Shoot-Look Shoot: A Review and Extension", *Operations Research*, vol. 52, pp. 454–463. [3.2, 3.5]

Grubbs, F. 1968. "Expected Target Damage for a Salvo of Rounds with Elliptical Normal Delivery and Damage Functions", *Operations Research*, vol. 16, pp. 1021–1026. [2.3.3]

Hammes, T.X. 2006. "Countering Evolved Insurgent Networks", *Military Review*, vol. 86, July–August 2006, pp 18–26. [10.4]

Hartmann, G. 1979. *Weapons That Wait*, Naval Institute Press, Annapolis. [8.1]

Helmbold, R. and Rehm, A. 1995. Translation of "The Influence of the Numerical Strength of Engaged Forces in Their Casualties", by M. Osipov, *Naval Research Logistics*, vol. 42, pp. 435–490. [5.2]

Hohzaki, R. and Washburn, A.R. 2001. "The Diesel Submarine Flaming Datum Problem", *Military Operations Research*, vol. 6, pp. 19–30. [7.6.3]

Hohzaki, R. 2007. "Discrete Search Allocation Game with False Contacts", *Naval Research Logistics*, vol. 54, pp. 46–58. [7.7]

Howes, D. and Thrall, R. 1973. "A Theory of Ideal Linear Weights for Heterogeneous Combat Forces", *Naval Research Logistics*, vol. 20, pp. 645–660. [5.5]

Hughes, W. 2000. *Fleet Tactics and Coastal Combat* (2nd ed.), Naval Institute Press, Annapolis, pp. 27–29. [5.1]

ICAO. 2003. *International Aeronautical and Maritime Search and Rescue Manual*, volume 2, International Civil Aviation Organization document 9731-AN/958, Appendix N, published jointly with the International Maritime Organization. [7.3.3]

Kaplan, E.; Craft, D.; and Wein, L. 2002. "Emergency response to a smallpox attack: the case for mass vaccination", *Proceedings of the National Academy of Sciences*, vol. 99 (16), pp. 10935–10940. [10.3.2]

Kaplan, E. and Kress, M. 2005. "Operational Effectiveness of Suicide Bomber Detector Scheme: A Best-Case Analysis", *Proceedings of the National Academy of Sciences*, vol. 102, pp. 10399–10404. [10.2]

Kaufman, H. and Lamb, J. 1967. "An Empirical Test of Game Theory as a Descriptive Model", *Perceptual and Motor Skills*, vol. 24, pp. 951–960. [6.3]

Keeley, R. 2003. "Understanding Landmines and Mine Action", http://www.minesactioncanada.org/techdocuments/UnderstandingLandmines _MineAction.pdf, accessed November, 2008. [8.1]

Kendall, M. 1959. "Hiawatha Designs an Experiment", *The American Statistician*, vol. 13, pp. 23–24. [2, introductory quote]

Kierstead, D. and DelBalzo, D. 2003. "A Genetic Algorithm Applied to Planning Search Paths in Complicated Environments", *Military Operations Research*, vol. 8, pp. 45–60. [7.5]

Kolmogorov, A. (ed.). 1948. "Collection of Articles on the Theory of Firing", Translation T-14, RAND Corp, Santa Monica. [2.3.2, 2.5]

Kooharian, A.; Saber, N.; and Young, H. 1969 "A Force Effectiveness Model with Area Defense of Targets", *Operations Research*, vol. 17, pp. 895–906. [4.3]

Kress, M. 2005a. "The Effect of Crowd Density on the Expected Number of Casualties in a Suicide Attack", *Naval Research Logistics*, vol. 52, pp. 22–29. [10.2]

Kress, M. 2005b. "The Effect of Social Mixing Controls on the Spread of Smallpox – A Two-Level Model", *Health Care Management Science*, vol. 8, pp. 277–289. [10.3.2]

Kress, M. 2006a. "Policies for Biodefense Revisited: The Prioritized Vaccination Process for Smallpox", *Annals of Operations Research*, vol. 148, pp. 5–23. [10.3.2]

Kress, M.; Baggesen, E.; and Gofer, E. 2006b "Probability Modeling of Autonomous Unmanned Combat Aerial Vehicles (UCAVs)", *Military Operations Research*, vol. 11, pp. 5–24. [9.4.2]

Kress, M. and Royset, J. 2008 "Aerial Search Optimization Model (ASOM) for UAVs in Special Operations", *Military Operations Research*, vol. 13, pp. 23–33. [9.4.1]

Lalley, S. and Robbins, H. 1987. "Asymptotically Minimax Search Strategies in the Plane", Proceedings of the National Academy of Sciences, vol. 84, pp. 2111–2112. [7.3.2, 7.6.3]

Lanchester, F. 1916. *Aircraft in Warfare: the Dawn of the Fourth Arm*, Constable, London. [5.2]

Lucas, T.W. 2000. "The Stochastic Versus Deterministic Argument for Combat Simulations: Tales of When the Average Won't Do!" *Military Operations Research: Special Issue on Warfare Analysis and Complexity-the New Sciences*, vol. 5, pp. 9–28. [1.3.1]

Lucas, T. and McGunnigle, J. 2003. "When is Model Complexity Too Much? Illustrating the Benefits of Simple Models with Hughes' Salvo Equations", *Naval Research Logistics*, vol. 50, pp. 197–217. [5.2]

Lucas, T., and Turkes, T. 2004. "Fitting Lanchester Equations to the Battles of Kursk and Ardennes", *Naval Research Logistics*, vol. 51, pp. 95–116. [5.4.2]

Lynn, J.A. 2005. "Patterns of Insurgency and Counterinsurgency", *Military Review*, July–August 2005, pp. 22–27. [10.4]

Manor, G. and Kress, M. 1997. "Optimality of the Greedy Shooting Strategy in the Presence of Incomplete Damage Information". *Naval Research Logistics*, vol. 44, pp. 613–622. [3.5]

Marcum, J. 1950. "Table of Q-Functions", report RM-339, RAND Corp., Santa Monica. [2.2.2]
Matlin, S. 1970. "A Review of the Literature on the Missile Allocation Problem", *Operations Research*, vol. 18, pp. 334–373. [4.3]
Matlin, S. 1972. "Equivalent Payload Nomogram", *Operations Research*, vol. 20, pp. 1190–1192. [2.3.1]
McCue, B. 1990. *U-Boats in the Bay of Biscay*, National Defense University Press, Washington, DC. [5.2]
McCurdy, M. 1987. "A Cognitive Planning Aid for Naval Minesweeping Operations", USCINCPAC Plans and Policy Directorate Research and Analysis Division, technical report. [8.4]
Miercourt, F. and Soland, R. 1971. "Optimal Allocation of Missiles Against Area and Point Defenses", *Operations Research*, vol. 19, pp. 605–617. [4.3]
Monach, R. and Baker, J. 2006. "Estimating Risk to Transiting Ships Due to Multiple Threat Mine Types", *Military Operations Research*, vol. 11, pp. 35–47. [8.4.4]
Morse, P. and Kimball, G. 1950. *Methods of Operations Research,* Technology Press. [1.3.4, 2.3.2, 2.5, 5.2, 5.3, 6.2.1]
Nadler, J. and Eilbott, J. 1971. "Optimal Sequential Aim Corrections for Attacking a Stationary Point Target", *Operations Research*, vol. 19, pp. 685–697. [3.3]
National Research Council, Naval Studies Board, Committee for Mine Warfare Assessment. 2001. *Naval Mine Warfare: Operational and Technical Challenges for Naval Forces*, National Academy Press, Washington, DC, p. 20. [8.1]
NSARC (National Search and Rescue Committee), http://www.uscg.mil/hq/g-o/g-opr/nsarc/nsarc.htm, IAMSAR supplement dated 2000, last accessed November, 2008. [7.2]
Odle, J. 1977."Minefield Analysis for Channelized Traffic", NSWC/WOL TR77-109, Naval Surface Weapons Center, August. [8.3]
OEG. 1946. "Search and Screening", report 56 of the Operations Evaluation Group, Department of the Navy. [7.2,7.3.3, 7.4,7.6.1]
O'Hanlon, M. and Campbell, J. 2007. *Iraq Index, Tracking Variables of Reconstruction Security in Post-Saddam Iraq*, The Brookings Institute, Washington, DC, July. [10.1]
Park, S. and Miller, K. 1988. "Random Number Generators: Good Ones are Hard to Find", *Communications of the ACM*, vol. 31, pp. 1192–1201. [Appendix C]
Perla, P. 1990. "The Art of Wargaming", Naval Institute Press, Annapolis. [6.3]
Pollitt, G. 2006. "Mine Countermeasures (MCM) Tactical Decision Aids (TDAs), a Historical Review", *Military Operations Research*, vol. 11, pp. 7–18. [8.4]
Puterman, M. 1994. *Markov Decision Processes*, Wiley, New York. [3.4.1]
Przemieniecki, J. 2000. *Mathematical Methods in Defense Analyses* (3rd ed.), American Institute of Aeronautics and Astronautics, Reston. [2.5,3.4.2, 4.2.1, 4.3]

Rasmussen, S. and Shima, T. 2008. "Tree Search Algorithm for Assigning Cooperating UAVs to Multiple Tasks", *International Journal of Robust Nonlinear Control*, vol. 18, pp. 135–153. [9.4.1]

Ravid, I. 1989. "Defense Before or After Bomb-release Line", *Operations Research*, vol. 37, pp. 700–715. [4.2.3]

Read, W. 1958. "Tactics and Deployment for Anti-Missile Defense", Bell Telephone Laboratories, Whippany, New York. [4.2.4]

Redmayne, J. 1996. "Evaluation of Mine Threat", Report SR-251, NATO SACLANT Undersea Research Centre. [8.4]

Richardson, H. and Stone, L. 1971. "Operations Analysis During the Underwater Search for the Scorpion", *Naval Research Logistics*, vol. 18, pp. 141–157. [7.5.1]

Robe, Q.; Edwards, N.; Murphy, D.; Thayer, N.; Hover, G.; and Kop, M. 1985. "Evaluation of Surface Craft and Ice Target Detection Performance by the AN/APS-135 Side-Looking Airborne Radar (SLAR)", Report CG-D-2-86, U.S. Department of Transportation, Washington, DC. [7.2]

Ross, S. 2000. *Introduction to Probability Models*, Harcourt, New York, ch. 6. [5.3]

Ruckle, W. 1983. *Geometric Games and their Applications*, Pitman, London. [6.2.3]

Skolnik, M. 2001. *Introduction to Radar Systems* (3rd ed.), McGraw Hill, New York. [7.1]

Soland, R. 1987. "Optimal Terminal Defense Tactics when Several Sequential Engagements are Possible", *Operations Research*, vol. 35, pp. 537–542. [4.2.1]

Stone, L. 1975. *Theory of Optimal Search*, Academic Press, New York. [2.5, 7.1, 7.5.2, 7.7]

Swain, J. 2005. http://www.lionhrtpub.com/orms/surveys/Simulation/Simulation.html, updated survey of simulation software last accessed August, 2008.[Appendix C]

Taha, R. 2007. *Operations Research, an Introduction*. Prentice Hall, New York.[Appendix C]

Taub, A. 1962. *John von Neumann Collected Works*, vol. IV, Pergamon Press, Oxford, pp. 492–506. [2.2.3]

Taylor, J. 1983. *Lanchester Models of Warfare*, INFORMS, Rockville, MD. [5.2, 5.3]

Thomas, C. 1956. "The Estimation of Bombers Surviving an Air Battle", The Assistant for Operations Analysis, Deputy Chief of Staff, Operations, Headquarters U. S. Air Force, Technical Memorandum 49, Appendix B. [2.4.4]

Thomas, C. and Deemer, W. 1957. "The Role of Operational Gaming in Operations Research", *Operations Research*, vol. 5, pp. 1–27. [6.3]

Tsipis, K. 1974. "The Calculus of Nuclear Counterforce", *Technology Review*, pp. 34–47. [2.3.1]

Turkes, T. 2000. "Fitting Lanchester and Other Equations to the Battle of Kursk Data", Masters Thesis, Naval Postgraduate School, Monterey, CA. [5.4.2, 5.5]

Urick, R. 1996. *Principles of Underwater Sound* (3rd ed.), Peninsula Publishing, CA.. [7.1]

USAF. 2005. http://www.eglin.af.mil/library/factsheets/, accessed November, 2008. [7.1]

von Neumann, J. 1928. "Zur Theorie der Gesellschaftsspiele", *Mathematische Annalen*, vol. 100, pp. 295–320. [6.2.1]

von Neumann, J. and Morgenstern, O. 1944. *Theory of Games and Economic Behaviour*, Princeton University Press, Princeton. [6.2, Appendix B]

Wagner, D.; Mylander, C.; and Sanders, T. 1999. *Naval Operations Analysis (ed. 3)*, Naval Institute Press, Annapolis. [4.2.1, 8.2]

Washburn, A. 1976. "Patrolling a Channel Revisited", Technical Report NPS55WS 75121, Naval Postgraduate School, Monterey, CA. [7.4]

Washburn, A. 1981. "Note on Constrained Maximization of a Sum," *Operations Research*, vol. 29, pp. 411–414. [7.5.2]

Washburn, A. 1995a. "Finite Method for a Nonlinear Allocation Problem," *Journal of Optimization Theory and Applications*, vol. 85, pp. 705–726. [2.4.3]

Washburn, A. 1995b. "MIXER: A TDA for Mixed Minefield Clearance", Project Report NPS-OR-95-011PR, Naval Postgraduate School, Monterey, CA. [8.4]

Washburn, A. and Wood, K. 1995. "Two-Person Zero-Sum Games for Network Interdiciton", *Operations Research*, vol. 43, pp. 243–251. [6.2.3]

Washburn, A. 2002. *Search and Detection* (4th ed.), INFORMS, Rockville, MD. [7.1, 7.3.3, 7.6.2, 7.6.3, 7.7]

Washburn, A. 2003a. "Diffuse Gaussian Multiple Shot Patterns", *Military Operations Research*, vol. 8, pp. 59–64. [2.3.3]

Washburn, A. 2003b. *Two-Person Zero-Sum Games*, INFORMS, Rockville, MD. [6.2.3]

Washburn, A. 2005. "The Bang-Soak Theory of Missile Attack and Terminal Defense", *Military Operations Research*, vol. 10, pp. 15–23. [4.3]

Washburn, A.R. 2006. "Continuous Network Interdiction," Technical Report NPS-OR-06-007, Naval Postgraduate School, Monterey, CA. [8.5.3]

Willard, D. 1962. "Lanchester as a Force in History: An Analysis of Land Battles of the Years 1618-l905," Research Analysis Corporation, RAC-TP-74. [5.4.2]

Winston, W. 1994. *Operations Research Applications and Algorithms* (3rd ed.), Duxbury Press, Belmont. [Appendix B]

Yost, K. and Washburn, A. 2000. "Optimizing Assignments of Air-to-Ground Assets and BDA Sensors," *Military Operations Research*, vol. 10, pp. 77–91. [3.5]

Zermelo, E. 1912. Über eine Anwendung der Mengenlehre auf die Theorie des Schachspiels. Proceedings of the fifth International Congress of Mathematicians, Cambridge 2, 501–510. [6.2.2]

Additional Reading

Bonder, S. 1967. "The Lanchester Attrition Rate Coefficient", *Operations Research*, vol. 15, pp. 221–232.

Grotte, J. 1982. "An Optimizing Nuclear Exchange Model for the Analysis of Nuclear War and Deterrence", *Operations Research*, vol. 30, pp. 428–445.

Shumate, K. and Howard, G. 1974. "A Proportional Defense Model", *Naval Research Logistics Quarterly*, vol. 21, pp. 69–78.

Soland, R. 1973. "Optimal Defensive Missile Allocation: A Discrete Min-Max Problem", *Operations Research*, vol. 21, pp. 590–596.

Swinson, G.; Randolph, P.; Dunn, B. and Walker, M. 1971. "A Model for Allocating Interceptors From Overlapping Batteries: A Method of Dynamic Programming", *Operations Research*, vol. 19, pp. 182–193.

USN. 1959. "Probability-of-Damage Problems of Frequent Occurrence", United States Navy, Operations Evaluation Group Study No. 626.

Weiss, H.K. 1953. "Methods for Computing the Effectiveness of Area Weapons", Ballistic Research Laboratory Report 879.

Index

Early Titles in the
INTERNATIONAL SERIES IN
OPERATIONS RESEARCH & MANAGEMENT SCIENCE
Frederick S. Hillier, Series Editor, *Stanford University*

Cox/ *RISK ANALYSIS OF COMPLEX AND UNCERTAIN SYSTEMS*
Williams/ *LOGIC AND INTEGER PROGRAMMING*
Saigal/ *A MODERN APPROACH TO LINEAR PROGRAMMING*
Nagurney/ *PROJECTED DYNAMICAL SYSTEMS & VARIATIONAL INEQUALITIES WITH APPLICATIONS*
Padberg & Rijal/ *LOCATION, SCHEDULING, DESIGN AND INTEGER PROGRAMMING*
Vanderbei/ *LINEAR PROGRAMMING*
Jaiswal/ *MILITARY OPERATIONS RESEARCH*
Gal & Greenberg/ *ADVANCES IN SENSITIVITY ANALYSIS & PARAMETRIC PROGRAMMING*
Prabhu/ *FOUNDATIONS OF QUEUEING THEORY*
Fang, Rajasekera & Tsao/ *ENTROPY OPTIMIZATION & MATHEMATICAL PROGRAMMING*
Yu/ *OR IN THE AIRLINE INDUSTRY*
Ho & Tang/ *PRODUCT VARIETY MANAGEMENT*
El-Taha & Stidham/ *SAMPLE-PATH ANALYSIS OF QUEUEING SYSTEMS*
Miettinen/ *NONLINEAR MULTIOBJECTIVE OPTIMIZATION*
Chao & Huntington/ *DESIGNING COMPETITIVE ELECTRICITY MARKETS*
Weglarz/ *PROJECT SCHEDULING: RECENT TRENDS & RESULTS*
Sahin & Polatoglu/ *QUALITY, WARRANTY AND PREVENTIVE MAINTENANCE*
Tavares/ *ADVANCES MODELS FOR PROJECT MANAGEMENT*
Tayur, Ganeshan & Magazine/ *QUANTITATIVE MODELS FOR SUPPLY CHAIN MANAGEMENT*
Weyant, J./ *ENERGY AND ENVIRONMENTAL POLICY MODELING*
Shanthikumar, J.G. & Sumita, U./ *APPLIED PROBABILITY AND STOCHASTIC PROCESSES*
Liu, B. & Esogbue, A.O./ *DECISION CRITERIA AND OPTIMAL INVENTORY PROCESSES*
Gal, T., Stewart, T.J., Hanne, T. / *MULTICRITERIA DECISION MAKING: Advances in MCDM Models, Algorithms, Theory, and Applications*
Fox, B.L. / *STRATEGIES FOR QUASI-MONTE CARLO*
Hall, R.W. / *HANDBOOK OF TRANSPORTATION SCIENCE*
Grassman, W.K. / *COMPUTATIONAL PROBABILITY*
Pomerol, J-C. & Barba-Romero, S. / *MULTICRITERION DECISION IN MANAGEMENT*
Axsäter, S. / *INVENTORY CONTROL*
Wolkowicz, H., Saigal, R., & Vandenberghe, L. / *HANDBOOK OF SEMI-DEFINITE PROGRAMMING: Theory, Algorithms, and Applications*
Hobbs, B.F. & Meier, P. / *ENERGY DECISIONS AND THE ENVIRONMENT: A Guide to the Use of Multicriteria Methods*
Dar-El, E. / *HUMAN LEARNING: From Learning Curves to Learning Organizations*
Armstrong, J.S. / *PRINCIPLES OF FORECASTING: A Handbook for Researchers and Practitioners*
Balsamo, S., Personé, V., & Onvural, R./ *ANALYSIS OF QUEUEING NETWORKS WITH BLOCKING*
Bouyssou, D. et al. / *EVALUATION AND DECISION MODELS: A Critical Perspective*
Hanne, T. / *INTELLIGENT STRATEGIES FOR META MULTIPLE CRITERIA DECISION MAKING*
Saaty, T. & Vargas, L. / *MODELS, METHODS, CONCEPTS and APPLICATIONS OF THE ANALYTIC HIERARCHY PROCESS*
Chatterjee, K. & Samuelson, W. / *GAME THEORY AND BUSINESS APPLICATIONS*
Hobbs, B. et al. / *THE NEXT GENERATION OF ELECTRIC POWER UNIT COMMITMENT MODELS*
Vanderbei, R.J. / *LINEAR PROGRAMMING: Foundations and Extensions, 2nd Ed.*
Kimms, A. / *MATHEMATICAL PROGRAMMING AND FINANCIAL OBJECTIVES FOR SCHEDULING PROJECTS*
Baptiste, P., Le Pape, C. & Nuijten, W. / *CONSTRAINT-BASED SCHEDULING*
Feinberg, E. & Shwartz, A. / *HANDBOOK OF MARKOV DECISION PROCESSES: Methods and Applications*

Early Titles in the
INTERNATIONAL SERIES IN OPERATIONS RESEARCH & MANAGEMENT SCIENCE
(Continued)

Ramík, J. & Vlach, M. / *GENERALIZED CONCAVITY IN FUZZY OPTIMIZATION AND DECISION ANALYSIS*
Song, J. & Yao, D. / *SUPPLY CHAIN STRUCTURES: Coordination, Information and Optimization*
Kozan, E. & Ohuchi, A. / *OPERATIONS RESEARCH/ MANAGEMENT SCIENCE AT WORK*
Bouyssou et al. / *AIDING DECISIONS WITH MULTIPLE CRITERIA: Essays in Honor of Bernard Roy*
Cox, Louis Anthony, Jr. / *RISK ANALYSIS: Foundations, Models and Methods*
Dror, M., L'Ecuyer, P. & Szidarovszky, F. / *MODELING UNCERTAINTY: An Examination of Stochastic Theory, Methods, and Applications*
Dokuchaev, N. / *DYNAMIC PORTFOLIO STRATEGIES: Quantitative Methods and Empirical Rules for Incomplete Information*
Sarker, R., Mohammadian, M. & Yao, X. / *EVOLUTIONARY OPTIMIZATION*
Demeulemeester, R. & Herroelen, W. / *PROJECT SCHEDULING: A Research Handbook*
Gazis, D.C. / *TRAFFIC THEORY*
Zhu/ *QUANTITATIVE MODELS FOR PERFORMANCE EVALUATION AND BENCHMARKING*
Ehrgott & Gandibleux/ *MULTIPLE CRITERIA OPTIMIZATION: State of the Art Annotated Bibliographical Surveys*
Bienstock/ *Potential Function Methods for Approx. Solving Linear Programming Problems*
Matsatsinis & Siskos/ *INTELLIGENT SUPPORT SYSTEMS FOR MARKETING DECISIONS*
Alpern & Gal/ *THE THEORY OF SEARCH GAMES AND RENDEZVOUS*
Hall/*HANDBOOK OF TRANSPORTATION SCIENCE - 2nd Ed.*
Glover & Kochenberger/ *HANDBOOK OF METAHEURISTICS*
Graves & Ringuest/ *MODELS AND METHODS FOR PROJECT SELECTION: Concepts from Management Science, Finance and Information Technology*
Hassin & Haviv/ *TO QUEUE OR NOT TO QUEUE: Equilibrium Behavior in Queueing Systems*
Gershwin et al/ *ANALYSIS & MODELING OF MANUFACTURING SYSTEMS*
Maros/ *COMPUTATIONAL TECHNIQUES OF THE SIMPLEX METHOD*
Harrison, Lee & Neale/ *THE PRACTICE OF SUPPLY CHAIN MANAGEMENT: Where Theory and Application Converge*
Shanthikumar, Yao & Zijm/ *STOCHASTIC MODELING AND OPTIMIZATION OF MANUFACTURING SYSTEMS AND SUPPLY CHAINS*
Nabrzyski, Schopf & Wglarz/ *GRID RESOURCE MANAGEMENT: State of the Art and Future Trends*
Thissen & Herder/ *CRITICAL INFRASTRUCTURES: State of the Art in Research and Application*
Carlsson, Fedrizzi, & Fullér/ *FUZZY LOGIC IN MANAGEMENT*
Soyer, Mazzuchi & Singpurwalla/ *MATHEMATICAL RELIABILITY: An Expository Perspective*
Chakravarty & Eliashberg/ *MANAGING BUSINESS INTERFACES: Marketing, Engineering, and Manufacturing Perspectives*
Talluri & van Ryzin/ *THE THEORY AND PRACTICE OF REVENUE MANAGEMENT*
Kavadias & Loch/*PROJECT SELECTION UNDER UNCERTAINTY: Dynamically Allocating Resources to Maximize Value*
Brandeau, Sainfort & Pierskalla/ *OPERATIONS RESEARCH AND HEALTH CARE: A Handbook of Methods and Applications*
Cooper, Seiford & Zhu/ *HANDBOOK OF DATA ENVELOPMENT ANALYSIS: Models and Methods*
Luenberger/ *LINEAR AND NONLINEAR PROGRAMMING, 2nd Ed.*
Sherbrooke/ *OPTIMAL INVENTORY MODELING OF SYSTEMS: Multi-Echelon Techniques, Second Edition*
Chu, Leung, Hui & Cheung/ *4th PARTY CYBER LOGISTICS FOR AIR CARGO*
Simchi-Levi, Wu & Shen/ *HANDBOOK OF QUANTITATIVE SUPPLY CHAIN ANALYSIS: Modeling in the E-Business Era*
Gass & Assad/ *AN ANNOTATED TIMELINE OF OPERATIONS RESEARCH: An Informal History*

Early Titles in the
INTERNATIONAL SERIES IN OPERATIONS RESEARCH & MANAGEMENT SCIENCE
(Continued)

Greenberg/ *TUTORIALS ON EMERGING METHODOLOGIES AND APPLICATIONS IN OPERATIONS RESEARCH*
Weber/ *UNCERTAINTY IN THE ELECTRIC POWER INDUSTRY: Methods and Models for Decision Support*
Figueira, Greco & Ehrgott/ *MULTIPLE CRITERIA DECISION ANALYSIS: State of the Art Surveys*
Reveliotis/ *REAL-TIME MANAGEMENT OF RESOURCE ALLOCATIONS SYSTEMS: A Discrete Event Systems Approach*
Kall & Mayer/ *STOCHASTIC LINEAR PROGRAMMING: Models, Theory, and Computation*
Sethi, Yan & Zhang/ *INVENTORY AND SUPPLY CHAIN MANAGEMENT WITH FORECAST UPDATES*
Cox/ *QUANTITATIVE HEALTH RISK ANALYSIS METHODS: Modeling the Human Health Impacts of Antibiotics Used in Food Animals*
Ching & Ng/ *MARKOV CHAINS: Models, Algorithms and Applications*
Li & Sun/ *NONLINEAR INTEGER PROGRAMMING*
Kaliszewski/ *SOFT COMPUTING FOR COMPLEX MULTIPLE CRITERIA DECISION MAKING*
Bouyssou et al/ *EVALUATION AND DECISION MODELS WITH MULTIPLE CRITERIA: Stepping stones for the analyst*
Blecker & Friedrich/ *MASS CUSTOMIZATION: Challenges and Solutions*
Appa, Pitsoulis & Williams/ *HANDBOOK ON MODELLING FOR DISCRETE OPTIMIZATION*
Herrmann/ *HANDBOOK OF PRODUCTION SCHEDULING*
Axsäter/ *INVENTORY CONTROL, 2nd Ed.*
Hall/ *PATIENT FLOW: Reducing Delay in Healthcare Delivery*
Józefowska & Wglarz/ *PERSPECTIVES IN MODERN PROJECT SCHEDULING*
Tian & Zhang/ *VACATION QUEUEING MODELS: Theory and Applications*
Yan, Yin & Zhang/ *STOCHASTIC PROCESSES, OPTIMIZATION, AND CONTROL THEORY APPLICATIONS IN FINANCIAL ENGINEERING, QUEUEING NETWORKS, AND MANUFACTURING SYSTEMS*

**** A list of the more recent publications in the series is at the front of the book ****

MIX
Papier aus verantwortungsvollen Quellen
Paper from responsible sources
FSC® C105338

If you have any concerns about our products,
you can contact us on
ProductSafety@springernature.com

In case Publisher is established outside the EU,
the EU authorized representative is:
Springer Nature Customer Service Center GmbH
Europaplatz 3, 69115 Heidelberg, Germany

Printed by Libri Plureos GmbH
in Hamburg, Germany